D0849893

Second-Order Consequences

A Methodological Essay on the Impact of Technology

TECHNOLOGY, SPACE, AND SOCIETY

Series Prepared by the American Academy of Arts and Sciences

GENERAL EDITOR:
  Raymond A. Bauer, Harvard Graduate School of
    Business Administration

Second-Order Consequences: A Methodological Essay on the Impact of Technology, by Raymond A. Bauer with Richard S. Rosenbloom and Laure Sharp — and the Assistance of Others
Social Indicators, edited by Raymond A. Bauer
The Railroad and the Space Program: An Exploration in Historical Analogy, edited by Bruce Mazlish

Second-Order Consequences

A Methodological Essay on the Impact of Technology

RAYMOND A. BAUER

with

Richard S. Rosenbloom *and* Laure Sharp — *and the*
Assistance of Others

THE M.I.T. PRESS
Cambridge, Massachusetts, and London, England

Foreword

In the past few years, my Subcommittee on Science, Research and Development has studied a number of issues in science policy. The course of our hearings and investigations has not been highly structured or planned. Much of what we have done has derived from questions before the Congress generated by the magnitude of Federal research and development expenditures, the pace of technological change, and the shifting priorities among national goals. However, in retrospect, a common theme is obvious — that is, growing recognition of extremely complex, close interactions between science and society, technology and culture.

The recognition is mutual. The ivory tower has disappeared, and scientists are eager to describe their research as relevant. Administrators and opinion leaders in industry and government interpret their projects and problems in terms of scientific descriptions and available technology.

The day of simple decisions based on direct, localized, and immediate costs and benefits is past. The most difficult and important issues in science policy are those where the broad, long-term assessment of applied science is necessary for decision. These issues baffle the conventional market place, the governmental regulatory agency, and the legislative committee. Where technology assessments are central to political judgments, they must be resolved before the total question can be dealt with satisfactorily. If assessment is delayed, or incomplete, or incompetent, then public policy is inadequate. In short, secondary consequences are of primary interest!

The methodology of identifying and measuring these secondary consequences is just beginning to develop. One important principle has already emerged. It is the danger that assessment automatically may become negative in character. Air pollution episodes, off-shore drilling accidents, military sonic boom flights over fragile earth formations, electric power failures — all are dramatic, news-making events. They are useful in calling our attention to the dependence of society on both effective, reliable technology and a sensitive environment. There is a need to direct prospective assessment to detect these unwanted consequences.

At the same time, I believe an assessment process must be conscientiously objective — seeking all effects, benefits, and drawbacks. There is a great deal to be gained in learning of unexpected *desirable* consequences. Management decisions can proceed at an earlier date; science may be applied with greater speed, and more confidence.

Another important result of understanding derivative effects and the ways in which they occur is the restoration of a sense of control over technology. We must halt and reverse the present trend toward a polarization of science vis-à-vis society. These antagonisms are fueled by ignorance of secondary consequences.

Thus, the studies which Professor Bauer has assembled here are an important initial contribution to the development of technology assessment. The focus on the space program is appropriate for two reasons: first, what we can learn of the real and extensive impacts on society in the early years of this large-scale technology will go far in counteracting the charges of technological stuntsmanship and cries for obvious, direct "spinoffs"; and second, what we can learn of the impact process itself may lead to more purposeful channeling of technology into the emerging social overhead problems.

The analysis of secondary consequences is going to become an integral part of the research-development-application sequence. Assessment work will be directed by the performers and managers of applied science. The value framework against which technological consequences are judged will be erected by the social sciences, the arts and humanities, and ethical and religious forces in our civilization. The responsibility for assuring complete and timely assessments will fall to political institutions — and my present conviction is that the Congress must fill this role. That is why I am greatly interested in this book and the methodology which is its subject.

EMILIO Q. DADDARIO
First District, Connecticut

April 15, 1969
2330 Rayburn Office Building
Washington, D.C. 20515

Preface

The work on which this book is based began more than six years prior to the writing of this Preface. In retrospect it appears that we embarked at an early time on a type of venture that has come, by late 1968, to be called "technology assessment." "Technology assessment" is a term used to refer to the as yet generally unpracticed practice of attempting to anticipate in advance, or as quickly as possible, the full range of costs and benefits that will flow from a particular piece or system of technology and to set policy accordingly. It has been made popular by Congressman Emilio Daddario, Chairman of the House Subcommittee on Science, Research and Development of the House Committee on Science and Astronautics. The National Academy of Engineering and the National Academy of Sciences have organized an effort to advise the Congress on this problem. Varied related proposals to consider technology assessment abound in Washington.

It is reasonable to suppose that we may be at a profound turning point in the way we manage our affairs. In the past the study of the second and third and more remote order consequences of technological change was left to the social historians. By now they have told enough dreary tales. Our environment — both physical and social — is so threatened, and the rate of actual and impending technological innovation is so vast, that apparently we have decided to do something about it. It is intriguing, therefore, that in the early months of 1962 the Space Agency came to the American Academy of Arts and Sciences with this issue.

How does one carry out technology assessment? I suppose that at this stage the problem is akin to that of how one can eat an elephant. The only answer is that one must begin by biting the elephant. And, considering the magnitude of the task, it is difficult

vii

to argue that one place is better than another for the biting to start. And, after a considerable amount of biting has taken place, the elephant remains largely unscathed—I fear.

The problem, as we saw it, was to understand how one might anticipate and/or detect at any early stage the social impact of an enterprise so vast as the U.S. space program. What my colleagues and I have attempted in this book is to chronicle what we did, and to report what we think we have learned. It seems fruitless for me to try to evaluate this effort. It is clear that we had to do a lot of groping. Over time our groping, I hope, became less random. Beyond that, we are at the mercy of the reader's judgment.

In an undertaking such as this, one must rely heavily on the quality of the people involved. Here I express my admiration and gratitude to those with whom I worked. We were blessed with the guidance of those Fellows of the American Academy who served on the Committee on Space Efforts and Society and its successor, the Committee on Space. In particular, I am indebted to the wisdom and patience of Mr. Earl Stevenson, chairman of the Committee on Space, with whom it was such a pleasure to work. Lewis A. Dexter, our executive director at the beginning of our work, made invaluable contributions. His wide-ranging mind, and his almost infinite network of intellectual contacts, made it possible for us to get some sense of direction. Edward Furash, as executive secretary to the Committee on Space, kept our affairs in order and was a diligent and able counselor to members of the research team.

The members of the core research team were Edward Furash, Bruce Mazlish, Robert Rapoport, Richard Rosenbloom, and Evon Z. Vogt. Their success in working together in a cooperative and creative fashion has strengthened my belief that interdisciplinary research is indeed possible. The traces of their contributions will be found throughout this volume. I have tried to be diligent in acknowledging these contributions when they can be identified, but their impact transcends that which can be labeled with any degree of specificity. This is especially true of Robert Rapoport, who wrote several essays that profoundly shaped the way I thought about the topics in the middle portions of the book.

John Voss, now executive officer of the Academy, joined our group when we were in midstream and has been immensely helpful and supportive since becoming executive officer. Without his strong assistance I would have despaired of ever getting this book out.

Finally, Maria Wilhelm has been of enormous help, editorially, in the final stages of the manuscript. She, too, saved me from despair.

Here, I fear, the list of specific credits must stop simply because the number of people who have helped in making this book possible is almost infinite. Many staff people contributed. Many scholars from academia, the government, and business were members of study groups or conferences, contributed working papers and memoranda, all of which were useful and some of which were highly original and profound. Many of their names will appear from place to place in the text where I have drawn directly on their work. But there are many others whose contributions were less direct though not necessarily less valuable. To them, too, I extend gratitude.

A project such as this generates a great many valuable manuscripts. We had contemplated publishing a volume of working papers. The problems of choice and format proved insurmountable. However, many of these papers are referred to in the text, and interested scholars may obtain copies of most of them by writing to the office of the American Academy of Arts and Sciences.

RAYMOND A. BAUER

Contents

Second-Order Consequences

A Methodological Essay on the Impact of Technology

Introduction

Our society has reached a stage at which it can afford to care, and cannot afford not to care, about the effects of its actions on its citizens. Presumably we have always held that the actions of any public agency (and most private agencies) were *intended* to be benevolent. Yet, it was also felt that there was some limit to responsibility. If dams were built, someone had to move. Railroads might open up a new territory, but they also cut through someone's land. Pulp mills turn wood into paper, but they also pollute the air and water and offend the senses of sight, sound, and smell. While such malevolent second-order effects of well-intended enterprises were recognized and generally lamented, they tended to be regarded as "inevitable" (a label calculated to turn the gold of resistance into the lead of folly). It is a mark of our times that we no longer accept a conclusion that we can do nothing about the unwanted consequences of our actions, or that a criterion such as "progress" in technology is *per se* a justification for imposing unpleasantness on ourselves.

What has happened is not so much a shift in awareness but a shift in our feeling of confidence and responsibility in planning and guiding a wider range of our affairs. The noxious by-products of our advances have long been recognized. So have the beneficial by-products of our less desirable actions; the technological by-products of our wars have long been a matter of comment. Particularly in the past decade, the American public seems less willing that either of these sorts of second-order consequences should be left to chance. Just as it is not acceptable to bear the undesirable consequences, there is also a growing feeling that beneficial side effects can be managed more skillfully. This book is a product of such feelings, an attempt to articulate and analyze them, and an assessment of what can be done about them.

More concretely, this is a record of the efforts of a group of scholars to look at the space program from the point of view of its second-order consequences. When Congress wrote the legislation establishing the National Aeronautics and Space Agency it took explicit recognition of the sorts of concerns just described and enjoined the new Agency toward

. . . the establishment of long-range studies of the potential benefits to be gained from, the opportunities for, and the problems involved in the utilization of aeronautical and space activities for peaceful and scientific purposes.[1]

From early in its history NASA was mindful of this mandate in its broadest sense. It commissioned a number of studies to explore the range of potential implications of its activities for areas of the society near and far from its primary mission. The most extensive of these early studies was done for The Brookings Institution by a study group under the direction of Dr. Donald Michael, eventuating in a book of several hundred pages.[2]

This concern of NASA with the wider ramifications of its actions was paralleled by public concern of a broader nature, which considered not only potential second-order consequences — both beneficial and nonbeneficial — but evaluated the space program itself in terms of these considerations along with its cost and its primary mission. The financial cost of the program, at the possible expense of highly valued goods such as education, housing, roads, and so on, had been a continuing concern of the general public.

In April of 1962 NASA, aware of these concerns, made a grant to the American Academy of Arts and Sciences

. . . for the support of basic scientific research entitled: Conduct a study of long-range national problems related to the development of the NASA programs.

Under this broad rubric the American Academy was asked to look at how

---

[1] National Aeronautics and Space Act of 1958, Section 102, c (4).

[2] Donald N. Michael, *Proposed Studies on the Implications of Peaceful Space Activities for Human Affairs* (Washington, D. C.: U.S. Government Printing Office, 1961). Available from Universal Microfilm Service, Inc., Ann Arbor, Mich., OP 17840, on microfilm or in standard journal size.

. . . the nation's resources may be mobilized for the achievement of national goals developing from advances in science and technology, with emphasis on those goals and advances as may be particularly related to NASA objectives; and of the effects of enterprises based on such mobilization of various sectors of society and NASA.

Stripped of the inelegance, which seems invariably linked to the language of contracts and grants, what the American Academy was asked to do was to consider how the second-order consequences of such large technological enterprises as the space program might be studied with respect to (1) making the beneficial effects more beneficial, and (2) making the malevolent effects less malevolent. This is a topic of great magnitude, ill defined, and one with which no one was or is particularly well equipped to deal. This is not to say it is a topic on which no one was ready to express opinions, but rather one on which the firmness of opinion was not matched by firmness of evidence. The strategy which the Space Committee of the American Academy developed was *to improve our ability to handle problems such as this, rather than either (1) speculate on the full scope of the issue, or (2) concentrate on one or a few substantive studies which in and of themselves would be important.*

Concentration on one or two or even three topics seemed to offer little opportunity to make a contribution to what might broadly be called the *management* of second-order consequences. Broad recognition that there are secondary effects resulting from technological change has been colored by a feeling of the inevitability rather than the manageability of these effects. The range of topics we proposed to study, in terms of potential management, was very broad indeed. To give some perspective on the scope, the problem, as we saw it, I will outline the topics which the NAS/NRC (National Academy of Science/National Research Council) study group suggested in the summer of 1962.

I. The Potential Economic Contribution of Space Expenditures
   1. Multiplier effect on GNP
   2. Multiplier effect at regional or local level
   3. Transfer of space technology to general economy
   4. Analysis of cost in a one-customer market
II. Manpower Resources
   1. Gross figures of scientific and engineering talents required
   2. NASA-University relations

      3. Retraining needs and possibilities
      4. Untapped potential of special groups
III. Social and Psychological Implications
      1. Impact on patterns of thought
      2. Public support of the space program
      3. The consequences of "Big Science"
IV. Space and Politics
      1. Space and foreign policy
      2. Cooperation with the Soviet Union
      3. Control of armaments and space
      4. Space experiments with potentially harmful effects
      5. Need for a coherent national strategy
      6. Other international implications

This list of topics is illustrative of the *potential* agenda with which we were confronted. From such a broad range of possibilities we had to formulate a task suitable to our resources and the times.

The decision to concentrate on an attempt to advance the art of managing second-order consequences suggested a three-pronged approach. The three prongs may be labeled Problem, Product, and Process.

The Problem

We took a step back to look at the over-all task of anticipating and measuring second-order consequences of actions. This involved the development of devices for anticipating a reasonable range of possible consequences in advance so that one might be prepared for them, as well as the development of a means for continually reassessing the probabilty of their occurrence. It also meant that if one were to take seriously the problem of social change resulting from such second-order consequences, he ought to have adequate measures of such change.

Early in our deliberation of this broad task, two key issues arose apropos the problem. The first was that of the utility of historical analogy as a device for generating a repertoire of probable consequences of large-scale technological change. The usefulness of such analogies was and is the subject of much debate. Our reply was to attempt an exploration in depth of one historical event, the early history of the railroads, as a potential broad analogue to the development of space exploration. The first volume of this series, *The*

Railroad and the Space Program, edited by Bruce Mazlish, reports the results of our study.[3]

The second key issue was that of the over-all status of our measures of social change. If we anticipate that something may happen, or think it has happened, we ought to be able to ascertain whether or not it in fact has happened. Yet, for a wide range of social phenomena, including many of those we take most seriously, we have poor measures, or none at all. This state of affairs precipitated an effort to review the status of our measures of social change, and to make a proposal as to the sort of societal information system needed in order to make the inferences that seemed called for if we are to take the planning and control of second-order consequences seriously. The result was the second volume in this series, *Social Indicators.*[4]

In the current (and third) volume I have concentrated on the problem of managing the consequences of technological change in a broad sense, rather than on specific consequences of the space program. In this situation as in others we found it impossible — or, if possible, undesirable — to draw a clear line between the space program and large-scale modern technology in general. Methodologically, the job of anticipating and measuring the social consequences of space exploration are those of any large-scale technological development. Of course there are aspects of the problem peculiar to the space program. But the art of managing second-order consequences is such that unless it is dealt with first, there would be little to be gained by concentrating solely on the impact of space exploration.

The Product

If an effect exists, it ought to exist in some identifiable and describable form. At the time our work began there was much speculation over what the effects of the space program would be or in fact already were — on people's imaginations, on support for science and technology, on attitudes toward religion, on communities near the aerospace industry, and on a variety of other things, including, of course, the impact of aerospace technology on the civilian economy. It occurred to us that the relative balance of speculation and

[3] Bruce Mazlish, ed., *The Railroad and the Space Program: An Exploration in Historical Analogy* (Cambridge, Mass.: The M.I.T. Press, 1965).

[4] Raymond A. Bauer, ed., *Social Indicators* (Cambridge, Mass.: The M.I.T. Press, 1966).

data ought to be redressed in some measure. As a result we undertook a series of essentially pilot investigations of the impact of the space program, and the modern technology of which it was the symbolic cutting edge, on special groups. The conclusions which I draw from these explorations in detecting and diagnosing the actual effect of the space program on such groups are presented in Part II of this book.

The aim of looking at a range of effects on a pilot basis was to discover answers to questions such as these:

1. Does preliminary exploration indicate that something beyond the self-evident might be found? For example, quite contrary to the expectations of many, religious fundamentalists do not seem concerned with the prospect of finding life in outer space. Or, the attitudes of businessmen toward the space program remained strongly positive at a point in time when the press suggested that there was growing, widespread criticism. (Regrettably, businessmen are the only segment of the public on which we had adequate trend data.)

2. Can the effects of the space program be separated from the effects of broad, massive, new technology?

3. Can a small-scale preliminary exploration suggest the directions that further research should take?

While our studies of "the product" of the impact of space exploration have substantive merit in their own right, their long-term contribution ought to be toward the illumination of what research on technological impact should be like — even though they be taken as bad examples.

The Process

Assuming technological change will have an impact, there must be some process whereby this impact is effected. How does technological information get from here to there? If one is to impede undesirable consequences and facilitate desirable ones, it will presumably be helpful to understand that process. Among the various side effects of space exploration discussed in recent years, one, the transfer of aerospace technology to the civilian economy, presented an especially attractive opportunity. In the early 1960's, when our work began, it was largely taken for granted that space technology would "spin off" or "drop out" automatically and quickly to the more prosaic sectors of the economy. This transfer was envisioned as one of the clear-cut major benefits to be anticipated from the space program.

We questioned this assumption, especially that the transfer would be automatic. The existence of the issue gave us a topic on which to undertake an exercise in studying the process whereby such a large technological event has wrought its effects. Since the beginning of our own efforts along this line, a number of other persons have also questioned the assumption that transfer takes place rapidly via natural processes, and I believe it can be said that not only has general knowledge of the process of technology transfer been improved, but the ability to manage the process has been demonstrated, at least to some extent.

The major consequence of our own work in this area is the research which has been carried on by Professor Richard Rosenbloom, whose essay on this topic is included in Part III of this volume. The work of Rosenbloom and his associates has been reported in various other places, and — as a good process should — has continued as an ongoing program of research under various auspices.

The remainder of this chapter will consist of a brief history of our activities. It will be followed by three sections: Problem, Product, and Process. In the first of these I will discuss the methodological issues as we see them, including the role of historical analogy, the need for measures of social change, and so on. The second of these sections will give an overview of our studies of the actual impact of the space program on special groups. This section will include one chapter in which our approach to a particular problem — the role of technicians in the manpower picture — is selected as an example in depth. The third section will deal with the process of impact in the area of technology transfer. In the convention of most books, I will then attempt a summary statement.

The Development of Our Research Project

In the spring of 1962, the American Academy of Arts and Sciences received from NASA a grant which eventually was to total $226,000. The purpose of this grant was to explore the social impact of the space program, the various ways in which such a large technological enterprise might affect our lives. To get this project under way the Academy appointed a Committee on Space Efforts and Society, composed of men of diverse backgrounds and experience, most of whom, however, had been active in public affairs. The members of that Committee were Raymond A. Bauer, chairman, Francis M. Bator, Saville R. Davis, Donald G. Marquis, Don K. Price, Walter A. Rosenblith, Earl P. Stevenson, and Arthur E. Sutherland. Lewis A. Dexter was executive director.

This Committee, as an over-all policy guiding group, quickly commissioned a Summer Study Group of qualified scholars to review the work that had been done on this and related topics, and to present specific proposals for studies to be conducted under the American Academy grant. In addition, the Committee surveyed a large sample of Fellows of the Academy, selecting them from the Academy classifications of Philosophy and Theology, Political Science, Social Relations, History, and Economics. All Fellows elected in 1961 were also included in the survey. The Committee communicated to this group its perception of the NASA grant and asked for suggestions for topics to be studied. At the end of the summer of 1962, the Committee received a rich list of suggestions from the two sources.[5]

The replies of Academy fellows reflected a wide area of concerns. These included the proper mission of NASA; the magnitude of the costs of the space program, and the alternate uses to which the money might be put; the impact of NASA on U.S. manpower needs (usually pessimistic); the effect of NASA on basic research and pure science, and on the university community (partly pessimistic and partly optimistic); the impact of the space program on liberal arts and the humanities (generally apprehensive); the impact on the economy (mixed); the administrative problems of NASA (mixed); public acceptance of NASA's program (apprehensive); and historical analogies that might illuminate space exploration.

It is clear that some of these suggestions fell outside of the mandate of our grant but all were helpful in shaping our sense of what concerned a highly educated group of citizens not themselves associated closely with science and technology. (*Note:* the Academy members surveyed did not include scientists and technologists, those persons who would probably have the closest knowledge of and identification with the space program. While the choice was deliberate at the time, in retrospect it would appear to have

5 Cf. Geno Ballotti, ed., "An Analysis and Summary of Responses from Fellows of the American Academy of Arts and Sciences Concerning the Space Program as a Subject for Study and Thought," a document of The Committee on Space Efforts and Society of the American Academy of Arts and Sciences (Cambridge, Mass., 1962), 16 pp.; John R. Seeley, Bertram M. Gross, Sumner Meyers, Lewis A. Dexter, Edward E. Furash, "Space, Society, and Social Science," a report to The Committee on Space Efforts and Society of the American Academy of Arts and Sciences (Cambridge, Mass., 1962), 43 pp.

deprived us of suggestions based on greater enthusiasm for and knowledge of the space program.)

The Summer Study Group cast its net widely. Its report ranged from a list of general and useful propositions about the utility of social science research, and the complications of applying it to policy problems, to a proposal for an "anthropological quinquennium" — a five-year project to gather "evanescent historical and anthropological data" which might be fast disappearing and which might serve as a useful base line in studying social change. The documents prepared by this group, including its summary report, proved very useful and are reflected in much of the work that we eventually did. However, the entire package they proposed would have demanded about twenty times the money we had, two and a half times the time we had, and approximately five to ten times the manpower. It was clear again that we had been provided with a rich menu from which we had to choose a worthwhile and digestable set of projects.

With these and other inputs, and with continuous conferring with NASA officials, the Committee worked out a research approach and work proposal of its own.[6] The Committee's *A Statement of Mission and Work* is presented as an appendix to this book. Where relevant, portions of the texts or paraphrases thereof are included in the body of this book. The reader who is interested in our over-all attempt to match modest resources to a vast topic is invited to peruse this appendix. The document was presented to the Fellows of the Academy at a meeting in February of 1963 and subsequently became the working paper on which the activities of the Academy in this area were based.

With the development of *A Statement of Mission and Work,* the task of the original Committee to develop policy and give direction was completed. The efforts of the Academy were reorganized for the accomplishment of the actual work to be done. Raymond A. Bauer, who had been chairman of the original Committee, became chairman of the Work Group on Space, composed of a number of active social researchers: Edward Furash, Bruce Mazlish, Robert W. Rapoport, Richard Rosenbloom, Earl P. Stevenson, and Evon Z. Vogt. This Work Group, in turn, reported to a newly formed Academy Committee, called the Committee on Space to distinguish it from its predecessor. This group was composed of scholars whose

---

[6] "Space Efforts and Society: A Statement of Mission and Work," a document of The Committee on Space Efforts and Society of the American Academy of Arts and Sciences (Boston, January 1963), 38 pp.

specialties were the various disciplines to be represented in the work to be done, who therefore could appropriately review the quality of that work. They were Earl P. Stevenson, chairman, Raymond A. Bauer, Richard H. Bolt, Gordon S. Brown, Max F. Millikan, and Talcott Parsons.

The activities of the Work Group on Space extended from early 1963 to 1964. The period since that time has been spent in writing and preparing manuscripts for publication, of which this volume is one. As I will state at several points throughout this book, we undertook an *exercise* in the study of the second-order consequences of such a large enterprise as the space program. This means that we deal with what *would* be involved if one took all these considerations seriously. Proceeding in this fashion is one valuable method of inspecting the assumption that some agency, NASA or any other, *might* want to assume responsibility for a wide range of such consequences.

We would like to anticipate the result of such an exercise. It is that any agency could assume responsibility for only a limited range of the secondary effects. To do more than that would paralyze it. Perhaps more pertinently one might argue that the range of interests impinged on is so wide that we could not burden an action agency with the enormous need for impartiality that would be demanded — or, if it were to assume the burden, it would be unlikely that the impartiality would be accepted by all parties. Furthermore, the effects of the program of any one agency interact with those of others and create public problems as well as opportunities that are better handled in their own terms. For example, if NASA causes manpower problems, so do other agencies, and some or most of these problems should be treated as national *manpower* problems, not as NASA problems.

On the whole, we will conclude that the management of second-order consequences is only in part, perhaps in small part, the task of the individual agency. It is rather a broad public policy that should be handled by other instrumentalities, public and perhaps private, of planning, control, and amelioration.

PART I

The Problem: Anticipating and Measuring Effects

1

The Secondary Impact
of Technology

"Can mankind survive in the kind of world that man is, himself, creating?" Five intellectuals debated this subject yesterday and concluded that the answer to the large question is a probable "yes." They left open such details as the quality and circumstances of that survival.[1]

The date of the debate was April 20, 1965. The place was the Regency Hotel in New York City. The number of "intellectuals" was actually six as reported in the headline, rather than the five reported in the story.

One might ask quizzically how man can have survived so long and still not be certain that it is possible — or, at least, if it is possible under decent conditions. Actually the question may now be more sharply put than ever before. The fate of an individual, a group, a society, has been a matter of concern as long as man has kept a record of his history and found that there was no reassurance that "everything will turn out all right in the end." In recent years, however, the technical ability to destroy ourselves has increased to the point where indeed it makes sense to think of the fate of mankind as a whole.

We are faced with a supreme example of the sort of paradox that marks human affairs, where the source of hope and the source of despair are only opposite sides of the very same coin. We have mastered the secret of matter, and we have developed the capacity to escape from the force of the earth's gravity. Thus we have broken through what a few years ago would have seemed absolute barriers, absolute limitations, on what man can do. But likewise,

---

[1] "6 [*sic*] Experts Debate Survival of Man," *The New York Times*, April 20, 1965.

one of the things that nuclear energy and missiles make possible is an unparalleled level of destructiveness.

This is not a book on the perils of the nuclear age. We use this example only to illustrate how man creates problems as he solves them, and so that we may pose quickly and sharply the question of the extent to which man can guide his own destiny.

The Impact of Technical "Improvements"

Doubt as to the extent, if any, to which we can control our own destiny, stems from the fact that even if we did things that we regarded as inherently desirable, they might produce second- and third-order consequences that were undesirable and of greater ultimate magnitude than the direct consequences of the original action.

Technical developments almost invariably have the appearance of being "good things," because they invariably begin as a solution to some agreed-upon problem. The task of solving the problem is seen as "purely technical." The underlying assumption behind the phrase "purely technical" is that no "policy implications" are involved — in the sense that one does not have to look beyond the solution of the immediate task to consider further ramifications of one's action. It should be noted that the implication behind the phrase "purely technical" is that one has arrived at an antiseptic solution for a socially posed problem.

Technology is often viewed as the mainspring of social change, but initially technology is the source of solution to problems that are social in origin, and technology is itself a social product. There is sophistication but shortsightedness in regarding technology as *the* prime mover of society. This sophistication, however, is a step forward from regarding benign solutions as having no ramifications beyond the resolution of the initiating problem. It is highly probable that there are many events which can be evaluated from the purely technical point of view as to how well a single task is performed. More usually, however, a "purely technical" analysis is an incomplete one.

Lauriston Sharp[2] tells us what happened to a group of Australian aborigines, the Yir Yorunt, when well-meaning missionaries gave them steel axes in place of their stone axes. It was hoped that the Yir Yorunt would improve their standard of living with the use of steel axes.

[2] Lauriston Sharp, "Steel Axes for Stone Age Australians," in Edward H.

But stone axes played important functions in Yir Yorunt life beyond that of cutting woods. The men owned the stone axes, which were symbols of masculinity and of respect for elders. The borrowing of axes from fathers, husbands, or uncles was regulated by Yir Yorunt custom. Furthermore, the ax heads were acquired from other tribes in a bartering relationship that was associated with elaborate rituals and seasonal fiestas.

The steel axes were readily accepted because the Yir Yorunt were accustomed to adopting goods acquired from other peoples. Since they were more efficient, the necessary work was accomplished in less time. This increased efficiency did not produce an increase in economic prosperity but in leisure time for sleep, "an act they had thoroughly mastered."

The missionaries had distributed the steel axes to men, women, and children without discrimination. But the older men, having less trust of missionaries, were not as likely to accept the steel axes. Soon elders of the tribe, once highly respected, were forced to borrow steel axes from women and younger men. The previous status relationships were thoroughly upset, and trading relationships and attendant ceremonials were disrupted.

The steel ax was the central, but — probably, granting the complexity of social causation — not the sole cause of the disruptions of the Yir Yorunt's social order. The generic story, however, is one that can be repeated almost endlessly from the pages of recorded history and the records of anthropology.

What is most impressive about these accounts is that a seemingly unimportant "technical" change produces a chain reaction — the implications of which were not seen in advance or sometimes not understood for a considerable period after the reaction had begun; the march of events appears uncontrollable, perhaps not subject even to any slight measure of redirection.

Of course, the outcomes are not necessarily unpleasant. The railroads hastened the formation of large cities. About the time that the cities faced the dire necessity for intra-urban transportation to get these seas of humanity from place to place within the cities, the peak of railroad building was passed. This released into the labor market a pool of civil engineers just about in time to meet the demands for their services in the cities, which they themselves had caused to come into being.[3] Sometimes, man is lucky.

Spicer, ed., *Human Problems in Technological Change* (New York: Russell Sage Foundation, 1952), pp. 69–92.

[3] Thomas P. Hughes, "A Technological Frontier: The Railway," in Bruce

When events turn out well, we learn little. As with the railroad builders in the preceding paragraph, we are unlikely to study the event and worry about what *might* have happened. The examples that capture and stimulate the imagination, like the impact of steel axes on Yir Yorunt culture cited previously, are those in which a technical improvement, highly desirable for its own sake, produced a pattern of widespread repercussions, which continued to be felt for a long time afterward.

In our own time, the impact of the automobile has been commented on endlessly. The automobile is credited with changes in American mores, extending from stealing automobiles for joy rides, through a decline in sexual morality, to getting the family out of doors for vacations in good clean air far from the city smog for which, in turn, the automobile itself is largely responsible. We can also cite the development of the suburbs and shopping centers out of the downtown sections, the accompanying demise of the downtown department store and of the railroads, and even the atrophy of leg muscles of potential Olympic distance runners.[4]

Perhaps the most thoroughly traced-out effect of a technical improvement is that described by Linton when writing about the Besileo, a tribe living on Madagascar.[5] At one time, rice was cultivated by the "dry" method. There were, however, swamps in which wet rice cultivation, a more efficient method, was practiced. Then irrigation was introduced. This meant that wet rice cultivation, a technically advantageous method, could be, and was, widely adopted.

Cohesive joint family groups that lived together, and were needed for cultivating the dry fields, were broken up. An irrigated

---

Mazlish, ed., *The Railroad and the Space Program: An Exploration in Historical Analogy* (Cambridge, Mass.: The M.I.T. Press, 1965), Chapter 2; especially pp. 68–70.

[4] It is worth an aside to indicate how firmly some of these beliefs can persist in the face of evidence if they happen to support broader myths in which people have a strong investment. Such a broad myth is that the automobile has brought about the physical deterioration of American youth. Evidence cited is the fact that American runners fare better in sprints than in distance races. Yet the superiority of American sprinters is largely accounted for by the group that has benefitted least from affluence and automobiles — the Negro youth, who are better sprinters and much poorer distance runners than are white athletes.

[5] Cf. Ralph Linton, "Ethnology of Madagascar" and "Tanala," in Abram Kardiner: *The Individual and His Society* (New York: Columbia University Press, 1939), pp. 282 ff.

field could be tended by one household. Other households were forced to move to progressively remoter distances in the jungle as land suitable for wet rice near the village was taken up. New villages were formed. However, the joint family retained its religious significance and met together for ceremonial purposes. Whereas marriages had previously taken place mainly within the local group, now the distribution of the joint family throughout many villages led to intervillage marriages. A sense of tribal identity developed.

Meanwhile the new villages, being associated with land of continuing fertility as a result of wet rice culture, became more permanent and demanded upkeep and defense.

Thus the tribal organization grew in solidarity as it was held together by the widely distributed joint families and as demands for military defense grew. Eventually the old tribal democracy disappeared and a king was chosen.

Linton traces out many other changes in Besileo life which resulted from the shift from dry to wet rice culture. Slaves became of economic importance. A caste system developed with a king at the head followed by nobles, commoners, and slaves. Religious life was modified, marriage practices altered, and so on. Note that the changes that took place in the lives of the Besileo as a result of adopting wet rice culture cannot be classified as uniformly desirable or undesirable. All that they have in common is that they were consequences which were neither intended nor anticipated. It is not our intention to intimate even fleetingly that second-order consequences or technical innovation are necessarily undesirable. All we contend is that they occur and in the past have been largely unintended and unanticipated. They thus pose the problem of whether or not man can avoid being the *unwitting* beneficiary or victim of his own actions. Or, stated more positively, can he deliberately take advantage of them?

These examples are sufficient to illustrate the phenomenon. Now, what are we to make of it?

Second-Order Consequences

In the first instance, we must observe that societies are complex phenomena with a myriad of functional interrelationships of which possibly only a minority are understood. While we have been exposed to ever-increasing assertions that the modern world is constantly more complex and interdependent, it is clear that even the primitive societies just described are of sufficient complexity to give one pause.

Therefore, we may repeat what has already been said; namely, that any action, no matter how beneficient its purpose, has wide-ranging consequences beyond its primary intent. True, in many instances these second-order consequences may be no more substantial in their final effect than the ripples that fan out when we drop a stone in the water. But this is not much consolation; if even only a very few innovations have serious, wide-ranging consequences, society may be drastically changed.

We have already mentioned that second-order consequences can be beneficial. Perhaps we should clarify what we mean by "unintended consequences" and "unanticipated consequences." A second-order consequence may be "unintended" but "anticipated": "I knew it was going to happen, but I had no option except to go ahead." Furthermore, all "unanticipated" consequences are not undesirable: "If I had known what was going to happen, I would have done it sooner." Second-order consequences may also be intended *and* beneficial: "I saw we could get a bonus in the form of an assist for so-and-so if I did it the other way, so that's the way I did it." Hence, second-order consequences may be intended or unintended, anticipated or unanticipated, desirable or undesirable.

Briefly stated, the major task in control over our destiny is to make as many second-order consequences as possible intended, anticipated, and desirable; and reduce to a practical minimum those that are unintended, unanticipated, and undesirable.

Stating a problem, however, is a far cry from solving it. All of our accounts of the social processes that follow in the wake of innovations are written from a historical perspective and give the image of an inevitable instance when the historian poses the question: "What would have happened if . . .?" Generally speaking, any such concern with the options available at various times, and the alternate choices which might have been made, is regarded as "iffy" and not worthy of serious intellectual concern. Hence there is no established body of historical (or anthropological) writing directed to the question of how the course of history might have been changed.

The functionalist tradition in anthropology has, in the main, studied social change with the purpose of tracing out the interrelationship of the various elements in the society, and of stressing the complexities of functional relationships among the elements in even very simple societies. It is not logically incompatible with a functionalist position to search for functional alternatives, or for areas of loose coupling in a society, but such tasks have not risen very high on the work agenda.

Hence, the weight of the writings on this subject is discouraging. It is discouraging in part, not because of the state of the real world, but because of the way in which the real world has been reported by men who have set themselves a special set of tasks. Even if there were no selectivity in reporting the impact of innovation on society, discouragement of this order is a luxury no one can afford. Someone must address himself to the task.

We argue throughout this volume and the accompanying volumes (of which this is in part a summary and interpretation) that there are two essential factors which will aid us in improving the prospect that our actions will produce desirable results or that undesirable results may be moved against early and effectively. These factors are (1) an increased understanding of our society, so that we may anticipate as far as possible the full range of consequences of our actions; and (2) an ability to detect unanticipated consequences as rapidly as possible so that beneficent ones may be facilitated and malevolent ones acted against.

Present Concerns

Concern for second-order consequences of man's actions is not a new thing. However, this concern has accelerated rapidly in recent years: "Indeed there are those who contend that the galloping technical revolution is threatening to outrun the number of talented people necessary to nourish it, as well as the time needed to plan and direct its course with some degree of wisdom."[6] We attribute this to three circumstances: (1) an increased understanding of the indirect implications of science and technology, (2) an increase in the affluence of the society, and (3) the space effort itself.

The growth of knowledge and general interest in the over-all problem is attested to by the fact that in 1959 the National Science Foundation began publishing an annual survey entitled *Current Projects on Economic and Social Implications of Science and Technology*. The first report in 1959 included somewhat over 100 projects; the second in 1960 over 200; the fifth in 1963 over 300; the

[6] *Report of the Subcommittee on Science Research and Development of the Committee on Science and Aeronautics*, U.S. House of Representatives (Washington, D. C.: U.S. Government Printing Office, 1963), p. 1.

sixth in 1964 about 580; and the seventh in 1965 slightly over 600. These increases are regarded by the compilers as due mainly to increased scholarly interest in the topic and only in a minor degree to an extension of the reporting system.

This growth of interest can be attributed in part to our increased sophistication about the interrelationship of science, technology, and society. But it must also be regarded as a reflection of the affluence of the society. There is a stereotype is some quarters that material wealth inhibits contemplation. But serious contemplation, particularly serious contemplation of the future, is a luxury that can be indulged in only if one has the time for contemplation and can envision having sufficient resources that, at a minimum, permit him to choose among alternative courses of action on the basis of not only their immediate, but also their remote consequences. The man who has to worry whether or not there will be food on the table tomorrow does not become a conservationist.

Such affluence also permits the initiating parties to be sensitive to the complaints of those who will, or think they will, be adversely affected. Urban renewal plans have begun increasingly to provide for the people who are displaced by such programs. This can be attributed in part to a better understanding of the consequences of displacing people. But it must also be attributed to the fact that American society of the mid-60's has the resources to cushion the shock of displacement rather than to force those persons being displaced to absorb the full burden on themselves.

Our technology has reached the point where we can contemplate rejecting strictly technological advances if they promise to produce adverse social results. The proposed supersonic commercial passenger plane (SST) is probably the first such major innovation for which probable public reaction was thoroughly researched in advance.[7]

The rapid spread of sophisticated technology, particularly the automation of many work processes, has generated a series of alarms and prophecies. There have been at least two major results of the books and articles written on the subject. International Business Machines Corporation, which spawns much of this technology, has granted $5,000,000 to Harvard University to study the relationship of technology to society. And a Presidential Commission on Technology, Automation, and Economic Progress was established

---

[7] This statement is made from the perspective of my membership on the NAS/NRC Committee on the SST-Sonic Boom.

and has delivered a report.[8] It does not seem inconceivable that in the future we may have a deliberate national policy of phasing-in new technology so as to minimize the adverse social consequences. The slogan, "You can't stand in the way of progress," is no longer unchallenged. At a minimum the issue seems to have been raised as to whether or not "progress" may be defined in narrowly technical terms.

First, the space program plays a special role, that of the symbolic cutting edge of this technology. It is the program that the public most readily recognizes as representing advanced technology. Second, space research involves exploration of areas about which few people have hitherto dreamed seriously and thereby evokes much curiosity as to the benefits and dangers that will result. Our favorite story about the anxieties that may be aroused is that of the woman who repeatedly scolded Werner Von Braun: "Instead of playing around with that space, why don't you stay home and watch television like God intended you to."

Third, it is a very large program, probably perceived as more of a unitary phenomenon than our military program. By this I mean that the average person's associations to the word "military" involve everything from a foot soldier through tanks, destroyers, aircraft carriers, to nuclear fission and fusion, and missilery, whereas when he speaks of space, he thinks of missiles, which vary only as to their destination and as to whether they are carrying men and instruments or solely instruments. Hence, the space program has a relatively narrow over-all public impact. Finally, it is a discretionary program. We do not *need* to do it in the same imperative sense that we believe we need to be militarily armed. This has become increasingly true as we have overcome the Soviet advantage in the space race. Hence, we are likely to deliberate more over its desirability.

Perhaps there has been too little attention paid to the importance of the fact that the space program *is* discretionary. Certainly in some quarters and at some times, this has been explicitly realized. We are speaking, for example, of the comparisons which are made of what might be done with an equivalent sum of money devoted to education, culture, public welfare, and so on. The fact that the space program is so freely exposed to criticism and evaluation is in itself a tacit recognition of its discretionary nature. But, beyond

[8] Report of National Commission on Technology, Automation, and Economic Progress (Washington, D. C.: U.S. Government Printing Office, January 1966).

all this, we believe that it is the discretionary nature of the program, together with its novelty, that has stimulated such soul-searching sensitivity to its second-order consequences.

As was indicated in the Introduction, this sensitivity to second-order consequences was initially reflected in the legislation establishing the Space Agency when a provision was inserted that NASA should make continuing studies so as to facilitate the benefits flowing from its activity. Specifically, the legislation stipulates "establishment of long-range studies of the potential benefits to be gained from the opportunities for, and the problem involved in the utilization of aeronautical and space activities for peaceful and scientific purposes."[9]

A considerable degree of the emphasis on second-order consequences has concentrated on practical applications and so-called "spill-over" to the civilian economy. This topic will be dealt with at length in a later chapter. However, we mention the fact here to note that a great deal of such discussion has been aimed at justifying the discretionary spending of so many billions of dollars on space exploration. A Congressional report of 1960 begins with these words:[10]

This report has been undertaken for a special reason. It is to explain to the taxpayer just why so many of his dollars are going into the American effort to explore space, and to indicate what he can expect in return which is of value to him.

NASA, in turn, has sponsored many efforts to study the effect of the space program on our life here on earth beyond the study on which we are reporting. References to such studies will be made at various points in this book. However, the outstanding example from the point of view of the topic with which we are immediately concerned is the study carried out under the direction of Donald Michael by The Brookings Institution during the years 1960–1961.[11] That report consists of more than 200 pages of informed speculation about future benefits and difficulties, direct and indirect,

[9] National Aeronautics and Space Act of 1958, Section 102 (c).

[10] *The Practical Value of Space Exploration,* Report of the Committee on Science and Astronautics, U.S. House of Representatives, 86th Congress, 2nd Session, Pursuant to H. Res. 133, July 5, 1960, Union Calendar No. 929, House Report No. 2091 (Washington, D. C.: U.S. Government Printing Office, 1960), p. 1.

[11] Donald N. Michael, *Proposed Studies on the Implications of Peaceful Space Activities for Human Affairs* (Washington, D. C.: U.S. Government Printing Office, 1961). Available from Universal Microfilm Service, Inc., Ann Arbor, Mich., OP 17840, on microfilm or in standard journal size.

intended and unintended, of the space program. It is doubtful that any other innovation in history was subject to such a single intensive consideration of its potential consequences.

We have spoken of an increasing understanding and concern over the second-order consequences of innovation. The space program, because of its distinctive characteristics, has accelerated the growth of this concern. It is conceivable that one of the major results of the space program may be its contribution to this very self-consciousness about the indirect results of our actions.

The consideration in this chapter of the indirect impact of technical innovation, while sketchy and discursive will, we hope, set the stage for a discussion of our own work, what we did, why we did it, and what we found out about the possibilities of directing the course of the secondary effects of space technology on society.

2

The Task of Anticipation

Attitudes Toward Planning and Control

In many of the world's largest and most successful companies, a well-planned system of planning is recognized as indispensable to success. Several years ago the Stanford Research Institute concluded, "In the cases of both high-growth and low-growth companies, those that now support planning programs have shown a superior growth rate in recent years." In a more recent study of the thirteen fastest growing companies in the United States, the author concluded that among the six major factors of growth found in all the companies was their success in introducing long-range planning programs and involving all levels of management in the process.[1]

Planning and social control are scarcely ideological issues any more in the United States. Whereas economic and social planning and control were, a few decades ago, tightly associated with socialism, they have become almost routinely accepted by business today. In fact, business's own planning is based on the assumption that the government will not only plan and control but be largely successful in its efforts.

Transitions of this sort are usually clearly identifiable if one makes comparisons over a long time period, but difficult to mark if one looks for a specific time of change or a single cause. Causation is usually multiple, probably because more than one element in the situation has to change in order to bring about a profound shift such as that in the American attitude toward planning.[2] In this case the time period—post World War II—is relatively easy to identify. As for the causes, perhaps the most direct positive cause has been the effectiveness of the "new economics" in guiding the

[1] George A. Steiner, "Improving the Transfer of Government-Sponsored Technology," *Business Horizons*, 9, No. 3 (1966), 55–62.

[2] For an account of this planning atmosphere in the business community, see John K. Galbraith, *The New Industrial State* (Boston: Houghton Mifflin, 1967).

stable growth of the U.S. economy in the past two decades, demonstrating that planning and a free economy are not antithetical. Complementary to this was the discrediting of the Soviet model of planning and control, which was basically a simple projection of the planner's intentions onto the future, to be achieved almost solely by administrative controls. Both of the foregoing developments are consistent with a broad philosophical change in orientation toward control of physical systems represented by the word "cybernetics."[3]

The cybernetic model of system control is based on the assumption that one can achieve at best only some approximation of what he aims at. Therefore salvation lies in rapid detection of error and adjustment to correct for that error. The word itself comes from the Greek word for "steersman" or "helmsman," and reflects the constant adjustments of course that characterize the way in which the sensitive helmsman controls his ship.

This cybernetic model also makes it natural to think of the second-order consequences of one's actions. "Error" is not a rare phenomenon that occurs because of bad planning and inept control. It is the natural and inevitable feature of all purposive action. Ironically, this technologically modern view of planning and control gives respectability to the once derogated British doctrine of "muddling through."[4] The early Soviet model, just referred to, was based rather explicitly on ignoring second-order consequences of one's actions. Unwarranted side effects were permitted to accumulate until they became of sufficient urgency to earn the status of a primary objective in their own right.[5] The British doctrine of muddling through, however, assumed a much more complex pattern of social processes. It might be described as deciding generally where one wants to go, taking a step in that direction, reassessing the consequences, reassessing one's goals and methods, taking another step, and so on. The doctrine of muddling through and the modern view of planning and control are therefore more complex than the cybernetic model in one crucial respect. In addition to providing for the correction of error they also provide for the revision of one's goals.

[3] This discussion is closely parallel to that in Raymond A. Bauer, ed., *Social Indicators* (Cambridge, Mass.: The M.I.T. Press, 1966), pp. 6–8.

[4] Cf. Charles Lindbloom, "The Science of Muddling Through," *Public Administration Review*, 19 (Spring 1959), 79–88.

[5] An extensive discussion of this view of the early Soviet model of planning and control is to be found in Raymond A. Bauer, Clyde Kluckhohn, and Alex Inkeles, *How the Soviet System Works* (Cambridge, Mass.: Harvard University Press, 1956), Chapters 4 and 5.

But, while it is now accepted that security and prosperity flow from planning, it takes some minimal base of security and prosperity to plan and control. The achievement of a certain level of affluence gives us confidence in raising our sights. Affluence gives us the ability to attend to second-order consequences and, as suggested above, to defer some immediate advantage because of longer-range considerations.

Meanwhile, back at the firm: the increasing complexity of the technology involved in modern business, and of the organizational structure and process, together with a confidence in anticipating the state of the economy a few years in advance, has produced a time perspective in the planning and control of American business in which resources—large resources—are committed years, sometimes decades, ahead of their anticipated payout.[6] While this change in business planning is in part a consequence of effective governmental control of the economy, it is also a stimulus to further and more effective government planning and control (of the basic features of the economy, not necessarily of the business firm's actions in the economy).

Thus we have identified a constellation of circumstances associated with a more sophisticated model of planning and control than we had contemplated in the past. To a large extent they reenforce each other. It is difficult to say that one is cause and the other effect. In some situations we may regard as effect that which was cause in another context. This is how systems theory tells us the world is.

In the years immediately past, the entire view of planning has undergone a fundamental revision.[7] What is of interest is that the most recent views are in many ways identical with the point of view with which we began our work almost five years ago. My conversations with, and my reading of the works of the new "futurologists," indicate that they are almost certain to be dealing with a convergent phenomenon which in effect has resulted in the parallel invention of many of the same ideas by a number of persons.

So that we may keep the time perspective of our own view of the planning and control process clear, I shall quote from our own document:

[6] See Galbraith, *The New Industrial State,* for a popular description of this process.

[7] Cf. Andrew Kopkind, "The Future-Planners," *The New Republic, 64,* No. 8 (February 25, 1967), 19.

. . . the appropriate incorporation of second-order consequences into the conduct of affairs involves the following problems: (1) anticipation and (to the extent that they cannot be anticipated) early detection; (2) evaluation; and (3) action.[8]

The problems of action and evaluation we felt to be largely, but not entirely, outside our purview since we saw our task as mainly that of improving our general capacity to analyze policy situations in such a way as to help policy makers who must make evaluations and design courses of action to do so more effectively. The key terms in the statement above are those of "anticipation" and "detection." The following passage conveys our thinking on this point:

. . . we will direct our efforts primarily at the technical problems of anticipating and detecting the second-order consequences of massive technological innovation, particularly as exemplified in the civilian space program.

The purpose of detection is to provide "feedback" to some action agency. We introduce the term feedback here because of the general notions of planning, action, and control with which the term is associated. Recognizing that its use in the social sciences is generally somewhat different than its more precise meaning in electrical engineering where it was developed, we will define this term for our purposes as, "the detection and reporting of the consequences of one's actions with sufficient rapidity and in such a form as to be a guide to corrective action." The addition of the notions of time and that the form of detection and reporting shall be appropriate to conceivable remedial actions is an important addition to the mere idea of detection. What we are concerned with, then, is not merely academic enumeration of lamentable catastrophes and missed opportunities, but the detection of things about which something might be done, and which might be reported soon enough and in such a form that something might be done about them.

This, however, is looking far down the road toward the day when feedback mechanisms might be built systematically into the nation's space efforts. For practical purposes, we will be occupied with *detection*, while keeping these other considerations in mind.

The companion term "anticipation" is also broader than another word that might have been used, namely "prediction". The distinction we would make is that between a reasonably precise knowledge of the probability of a given event happening, and a sufficient awareness that . . .

---

[8] "Space Efforts and Society: A Statement of Mission and Work," a document of The Committee on Space Efforts and Society of the American Academy of Arts and Sciences (Boston, January 1963), p. 11; see Appendix, pp. 201–202.

if it does happen we are not caught by surprise. In planning it is often possible to make provision for handling some consequence even though it can be foreseen only in very general terms.

There is another sense in which the distinction between "anticipation" and "prediction" is vital. Prediction is an enterprise which places a primary emphasis on the probability of a given event occurring in the future. Prediction *per se* should not be concerned with the *importance* of various possible future events—though as a matter of fact the probability and the importance of the events often get confused in people's minds.[9]

If we succeed in keeping the issues straight, we may "anticipate" the possibility of some future event of great importance, though other outcomes might be more probable. Yet, its importance relative to other possible outcomes might be so great that we would be wise to act as if it were going to occur. Thus, when we embark on an automobile trip, the probability of encountering an accident is low, but this is no reason for not using a seat-belt. Or, viewing the matter in reverse, using a seat-belt does not constitute a prediction that one will have an accident.

This distinction takes on particular relevance in the light of the proclivity of social scientists and social critics to write "cautionary tales" such as *1984* or *Brave New World* as though they were the most probable estimates of the way things will be. They are often well warranted as "forecasts" of *possible* contingencies. But they should not be confused with "predictions" of what is most likely to be. It is an empirical question as to which question is more important, but in planning for the future the important but improbable outcome is often more relevant than the unimportant but probable one. "Anticipation" is a word that directs our attention to the full range of possible outcomes, while "prediction" directs our attention to the most probable.[10]

The "new" features of this approach are

1. Treatment of "detection" as an essential part of a feedback system. In the modern view of planning, detection is not used for historical purposes, or for a post mortem to guide one in undertaking future efforts, or even as an emergency feature to detect disaster, but rather as a matter-of-fact, essential, day-to-day feature of any system of action. It may be used to take advantage of opportunities, and it is assumed that it will be used to minimize minor problems as well as to avert disaster.

2. Emphasis on "anticipation." This has an even more modern sound. The modern French planner-theoretician, Bertrand de

---

[9] For a discussion of this issue see Raymond A. Bauer, "Accuracy of Perception in International Relations," *Teachers College Record*, 64 (January 1963), 291–299.

[10] "Space Efforts and Society," pp. 12–14; see the Appendix, pp. 203–205.

Jouvenel, has employed the term "possible futures" in labeling his planning project *Futuribles*. Furthermore, the phrase "inventing futures" has become a standard phrase among the recent "futurologists."

De Jouvenel's reasoning for advocating a consideration of a plurality of future states is very similar to our own.

What is meant by the speaker as the formulation of a very likely course is all too often understood by the listener as a statement of what is *bound* to happen. Hence the new superstition that we are slaves of a given future. . . . Now it is one thing to admit that such courses are extremely likely, and quite another thing to take them as necessary. . . . it is logically obvious that we could reverse these trends if we chose. . . .[11]

The modern view of planning might be summarized in these terms: Anticipate that range of future states that is reasonably probable and of importance to you. Make a strategic decision as to which of these states you would like to make more probable. (This is a difficult decision, since the most desirable state may be disproportionately hard to obtain.) Make constant readings to assess the changing probabilities of this set of future states. Reconsider your strategy and your goals.

The foregoing statement does not constitute a proposal on our part to take over the planning function of NASA or of any other agency but rather to put into perspective the need for considering the joint tasks of anticipation and detection of consequences.

We arrived at this position not through an overpowering drive to return to "first principles," but through trying to make sense out of that which was going on about us. We were confronted by statements of gloom and of glee as to what the secondary effects of the space program might be:

Man's conceptions of himself and of God would be affected, whether by demonstration of his new technological ability, or by discovering other potentially more intelligent beings in this or other universes. Satellite communications would improve T.V. programming, or solidify the grip of mediocrity. The space race with the U.S.S.R. would improve or worsen relations with the Soviet Union. Moment to moment surveillance of military installations throughout the world would increase or decrease tension. Improved weather forecasting would be a boon to mankind, or a source of embarrassment to the man who had to tell the bad news to the resort operators. Our economy would be stimulated, or drained. The

[11] Bertrand de Jouvenel, "Notes on Social Forecasting," a speech given in London, June 1966.

educational aspiration of school children would be raised by the example of the new technology, or lowered by a feeling of defeat in the face of overwhelming demands. The dignity of man would be raised, or lowered. The prestige of the U.S. would be raised, or lowered.[12]

How was one to sort such statements? What role did they have in a consideration of the second-order consequences of the space program? How was one to know if these things happened? What other things might happen that hadn't been mentioned? Were these supposed to be statements of caution, or premature justifications of the space program?

Devices of Anticipation

When we, as a first step in trying to answer these questions, took a look at some of the aspects of U.S. manpower problems as they might be related to the space program we ran into one government official of long standing and proved competence who complained that predicting manpower requirements would be quite simple "if some one would only tell me what assumption to make." Upon being queried as to what he meant by his complaint, he indicated that what he wanted was for the Secretary of Defense to tell him whether to make his projections on the assumption that military expenditures would remain the same, would rise, or would decrease. If he knew the impact of our military effort on the over-all manpower situation, he could then make a guess whether the space program would be helpful or harmful, with respect to manpower needs. Yet to use any one of these assumptions seems certainly the wrong thing to do. No one of these three states of affairs is certain to occur. If any one of them were close to certainty, it would be self-evident, and he would have had no occasion to call for guidance. Obviously the sensible way to plan for manpower needs would be to take into account all three states of affairs. (I speak of "three" states of affairs knowing full well that an increase or decrease in military expenditures represents two ranges of states. The reader may inject whatever level of complexity suits his taste.) The best prediction of future manpower needs would be one based on an

---

[12] Raymond A. Bauer, "Space Programs: The Joint Responsibility of Business and Government," an address to the Harvard Business School Alumni Association, Baltimore, Maryland, April 9, 1964.

estimate of the relative probability of each of the three (or more) states of affairs occurring in the future.

And, a plan properly drawn would provide for a continuing re-assessment of changes in the probability of military expenditures going up, dropping, or staying the same. A proper plan would include contingency plans for responding to changing situations.

All in all, appropriate planning is dependent on the number of relevant, plausible future states we can anticipate — or in more popular terms, "predict."

But how does one predict? Prediction of the future is an activity that man engages in as much as any other. It is also the one activity that is most often labeled "Impossible — Who can predict what is going to happen?" But, on the other hand, how can anyone avoid prediction? Behind every present action of ours is the assumption that at least it probably won't make things worse. If we were not reasonably content with that expectation we would ordinarily find something else to do.

This brings us back to the problem of the second-order consequences of technological change. How are we to generate a reasonable set of expectations as to the consequences that may flow from a given technical change? (The obvious answer is that we must have a "model.")[13] The common model used in practical forecasting, whether it be of economic growth, populations, technology, and so on, assumes that the phenomenon in question will continue to grow, probably at an exponential rate. However, this method of projecting the past into the future is not a particularly useful model for the task with which we are concerned. It may be that technology itself will grow at an exponential rate. But this tells us nothing about the qualitatively different sorts of events that technological growth will activate. There is a great deal of difference between projecting the rate of growth of a phenomenon for which we already have a series of historical data and anticipating the consequences of that growth.

The tools we have for making such anticipations are several. The most basic of such tools is our general conviction about the complexity of the social process, namely, that some second-order consequences *will* take place. This is an assumption of such great generality, and currently widely enough accepted, that it takes

[13] For a most comprehensive statement of modes of prediction, see Daniel Bell, "Twelve Modes of Prediction — A Preliminary Sorting of Approaches in the Social Sciences," *Daedalus, 93* (September 1964), 845–880.

special effort to remember it. Beyond this assumption that something will happen we have a number of concepts that help us restrict the range of things that may happen. Here I borrow from Bell's excellent essay on modes of prediction:[14]

There are certain *structural certainties* that characterize most future states. Bell paraphrases de Jouvenel's example of a U.S. presidential aspirant who belongs to the same party as the incumbent. The aspirant knows from the realities of the American political scene that the incumbent will almost certainly be nominated for a second term, but cannot by law be nominated for a third term. Hence, the aspirant knows that to time his candidacy properly he must avoid moving too fast, but must nevertheless be "ready" for the presidential election after the next. There are various other such forms of institutionalized behavior, some of which will have to be changed if it is inappropriate to the events that ensue.

Bell cites "The Operational Code" of a society and "The Operational System," both of which are more implicit than the structural certainties mentioned above, but which also share the characteristic of describing "how things are done" in a given society. All of these aspects of the situation serve to make some types of response more likely than others. They become relevant in reducing the uncertainty of our anticipations, but they can do so only after we know what class of events to start worrying about.

The functional prerequisites of the society are one feature of every human situation that gives a clue as to what sort of events to anticipate. Generalized functional prerequisites for all societies have been identified. They include the biological necessities of food, shelter, and reproduction, and the sociological ones such as social-control mechanisms. In addition, in any one section of a society certain features may come to be regarded as a person's "natural right," such as owning a TV set or an automobile. We may expect any violation of what people regard as their legitimate expectation to produce some counterreaction, even though these expectations may have nothing to do with the survival of the race.

The actual anticipation of a specific future state that may follow from a given technological change is a work of imagination. Into this work of imagination are fed the factual information that is available, one's generalized knowledge of the social institutions and the people with whom he is dealing, and any appropriate models or other stimuli to the imagination that may be available.

[14] *Ibid.*

The Use of Analogy

The most pervasive source of models or stimuli to the imagination appears to be analogy. In general, when we are confronted with a social situation of great complexity we tend to classify it with "similar" situations before we begin any careful analysis. The alternative might, in principle, be to do a thoroughgoing analysis of the situation in its own terms, bringing together whatever theoretical and factual tools we have to determine the outcome of the situation.

The fact is that for complex events such as the exploration of space, or even for much simpler societal events, we have no analytical model with which a comparison might be made. However, there is no doubt that the economist and sometimes even the sociologist can analyze some *portions* of the phenomenon. Given this situation, our tendency is to find some category to which to assign the event, or some other event that seems similar to it. For example, space exploration is seen as an instance of "exploration" or of "discovery," and it is similar to the finding of the New World (if it is "exploration"), or perhaps (if it is "discovery") to the impact of the microscope or the telescope.

Early in the operation of the American Academy's Committee on Space, we learned that the use of analogy was rampant in discussion of the space program. Most frequently, it was compared to the voyages of discovery or to the Copernican revolution and other such events that reputedly had changed man's conception of himself and his universe. Or, it was compared to previous technological or economic innovations that had been resisted by allegedly shortsighted people who could not envision the full benefits. Congressman George P. Miller, for example, quoted Daniel Webster on the proposals that the Government help open up the West — "What do we want with this vast, worthless area, this region of savages and wild beasts, of shifting sands and whirlpools of dust?" Mr. Miller uses this example to reach the following conclusion: "So, whoever says going to the moon and Mars is a waste of money is risking the vengeance of history."[15]

Close to historical analogies as a source of imagery for talking and thinking about the space program was science fiction, e.g.,

[15] Quoted in the Wall Street Journal, June 27, 1963; cited by Bruce Mazlish in "Historical Analogy: The Railroad and the Space Program and Their Impact on Society," in Bruce Mazlish, ed., *The Railroad and the Space Program: An Exploration in Historical Analogy* (Cambridge, Mass.: The M.I.T. Press, 1965), p. 1.

"Buck Rogers stuff." Early in our study efforts, we commissioned a number of exploratory essays on the use of both historical analogy and science fiction as devices of anticipation of the second-order consequences of space exploration.[16] Historical analogy eventually suggested itself as the more fruitful device. Science fiction is limited because by and large it deals solely with space flight and pays very little attention to those other activities associated with the space program, including the massive technological and economic effort behind space flight and space science. Equally serious, it practically ignores second-order consequences:

The writers, when they mention these things at all, usually posit a general public enthusiasm—sometimes spontaneous, sometimes manipulated . . .—but they have said little beyond this.[17]

Another reason was proffered by the authors of our memoranda, namely, that science fiction had not proved to have a very high predictive accuracy. Storer cited as a limitation of this genre that science fiction writers had predicted virtually everything. Coming very early in our work, this statement of limitation went unchallenged. As, over time, we came to realize that the task of planning does not call for precise prediction but, as I have said repeatedly, for broad-range anticipation, we might not have regarded this condition of science fiction as a limitation but as an asset. In any event, by the time we realized this, we had taken the road of historical analogy.

Our work on historical analogy was carried out under the direction of Professor Bruce Mazlish of M.I.T. The passages that follow are based on his volume, *The Railroad and the Space Program,* and especially on his introductory essay. I must take the responsibility for both the form and substance of these conclusions. Although they are dependent on the work of Mazlish and his coworkers, they are selected and phrased to suit the context.

The Status of Analogical Thinking

"It is extraordinary," wrote Mazlish, "how often historical analogies are used, and how little reflection is given to their usage." A review of the bibliography of works in the philosophy of history, published in *History and Theory* (Beiheft I) containing hundreds

---

[16] Cf. Norman W. Storer, "Modern Science Fiction: Its Potential Use in Predicting the Consequences of Space Exploration," a report to The Committee on Space Efforts and Society of the American Academy of Arts and Sciences (Cambridge, Mass., June 1962), 33 pp.
[17] *Ibid.,* p. 28.

of references to books and articles, located no reference to historical analogy.

And, while analogy is referred to in all textbooks on logic, it virtually never receives any full-scale treatment.[18] It is generally considered as "the most primitive, and, at the same time, one of the most important of all forms of reasoning." Because two events have enough in common to be seen as "similar" it is assumed that the less well-known event may share still other attributes of the better-known event. While we restrict our discussion here to historical analogy, it is clear that analogical thinking can be based on any perception of similarity between events that are otherwise in some way also different. Thus, the model of a machine with moving parts that have a determinate effect, one on the other, has been used for the study of human behavior, and especially for the study of human organizations. In turn, biological models have been used as analogies for social phenomena. However, regardless of their origins, analogies have the same functions and present the same problems.

To the extent that an analogy is used as a source of hypotheses about some contemporary event, its validity lies in whether or not those hypotheses are fruitful in giving us insight into that event. The danger in the use of analogy lies in failing to check its validity in the individual instance. It is one thing for Congressman Miller, cited previously, to look to the history of attitudes concerning the opening of the West as a source for the hypothesis that one *may* underestimate the value of an undertaking such as the space program. It is quite another thing to assume that because Webster underestimated the value of opening up the West, that anyone who questions the value of the space program is therefore necessarily wrong.

Matters might be somewhat different if we had a fair sample of such historical events, and a large enough sample that we could make a statement of this order:

The weight of historical evidence is that men will usually underestimate the benefits of broad large scale enterprises because such enterprises are evaluated solely in terms of their foreseeable direct benefits. While unforeseen benefits cannot, by definition, be foreseen, one can count on the number being large enough, and their importance great enough that they will constitute a sizable contribution to the total benefits.

It goes without saying that such evidence has not been gathered

[18] Mazlish found one, Harald Hoffding, *Der Begriff der Analogie* (Leipzig, 1924).

systematically and that there are few such general statements that can be made.

There are, therefore, several ways in which analogies may be used, each way presenting some utility and some potential dangers.

1. As a source of hypotheses to be validated against present events, there can be little quarrel with the use of analogies. They either give us fruitful insights, or they do not. While I do not care to espouse the writing of bad history, the historical soundness of the analogy is in this case not crucial. When the analogy is used as a source of hypotheses, "validation" lies in the current events. It may be taken for granted that its *usefulness* will be enhanced by the soundness of the historical work.

2. Historical analogies may be used as a source of general statements about social affairs. Here the issue of the methodology of the historical studies becomes central, a point that I shall discuss later. There is a great danger in confusing this and the preceding use of analogy. This latter danger will be discussed immediately.

3. Historical analogies adequate for generating hypotheses are often treated as though they were adequate for making general statements. These general statements then serve what Mazlish calls a "mythical function," as a guide to action. Two errors are committed here. First the validity of the generalization as a description of social phenomena is overestimated. Second, a general proposition is taken as a guide to specific action. Generalized knowledge improves our ability to predict (and here I use the term literally). However, in any individual situation we must take account of the idiosyncratic features which characterize that situation and modify any general process.

4. Analogy is also used as a means of communication. For example, President Kennedy referred to space as "a new ocean" on which "we must sail." And, as Mazlish points out, the language of the historian is riddled with analogy. When one talks of "the Space Revolution," or "the Industrial Revolution," says Mazlish, "he has borrowed his term, by analogy, from the original revolutions of the heavenly bodies."

Such analogies are useful ways of drawing on the shared experiences of the writer and the reader. Like all figures of speech, however, they have an access of semantic baggage and may be misleading. Given this assessment, how does one go about the business of studying analogies in a useful way?

The Contribution of Historical Analogy

The question of whether or not historical analogies are "valid" or "useful," put that simply, is not a very meaningful one. The usefulness or validity of an historical analogy must, as I have just pointed out, vary with the specific situation. However, the utility or validity of historical analogies had not, to our knowledge, been explored in any systematic fashion. Our reasoning was that thought or talk on the topic could be facilitated only if someone were to make such an effort.

I should say in advance that whether we consider this effort successful or unsuccessful, its major contribution appears to have been not so much in the suggestion of *specific* propositions as to what might be expected in terms of second-order consequences from the space program, but in a better conceptualization of what such a social event *is* and of how we ought to go about studying it.

Any event of this complexity may be seen as many things. The prevalent tendency, as mentioned previously, has been to think of the space program as something analogous to voyages of exploration or to intellectual discoveries that changed our perspective on our selves and our world. But, in Mazlish's terms it is more than this: it is an enormous "social invention," with technological, economic, sociological, political, and intellectual aspects and consequences. It was for this reason that he chose to study a social invention comparable in complexity, size, and relative importance to the economy and the society, namely, the development of railroads in America.

To some extent the historical studies of the railroad converged on generalizations that we arrived at when we studied the space program directly. For example, such large-scale social inventions take on dramatic and symbolic value such that they are seen as discrete entities with a relatively distinctive role in the society of their time. Upon close inspection they turn out to be complex entities, complexly related to the other events of their time, difficult to distinguish at many points, and probably having less impact than the dramatic and symbolic view of them suggests. Or perhaps one could say that one of their chief functions is to serve as a symbol and a source of myths around which larger societal efforts are organized.

Mazlish offers the following broad conclusions, which he frankly labels as "only a halfway step from generalities to generalizations."

A. All social inventions are part and parcel of a complex—and have complex results. . . . While this may seem like stating the obvious . . .

(it) should serve as a warning to anyone, scholar, publicist, or policy maker, to be cautious about embracing simple views as to the nature and impact of a social invention.

B. No social invention can have an overwhelming and uniquely determining economic impact, and this is so partly because no completely new innovation is possible in reference to any set of economic objectives. [Whatever the functions served by a given social convention there will be other events in the society that to some extent serve each of these functions.]

C. All social inventions will aid in some areas and developments, but will blight others.

D. All social inventions develop in stages, and have different effects during different parts of their development.

E. All social inventions take place in terms of a national "style," which strongly affects both their emergence and their impact.[19]

While the foregoing statements may strike the reader more as "generalities" than as "generalizations," each reflects concrete facets of the development of the railroads that had been overlooked, or not understood. For example, the railroad had been seen as the prime mover in the development of the economy of the mid-nineteenth century, and specifically of the development of certain industries, such as the steel industry. "The railroad was not 'the exogenous force which stimulated the pace of economic development,' but rather followed the pattern of general investment."[20]

In point of fact, the railroad appears to have followed economic development as much as it stimulated it. For the first several decades it stimulated the British rather than the American steel industry: British steel was used for the rails until the mid-century, when the railroads moved far enough inland that the cost of transportation offset the price advantage of British steel. However, since the American steel industry grew during the early history of the railroad, a causal relationship was easy to make and slow to be challenged.

As a matter of fact, the building of the railroads and their opening of new territory captured the attention of historians to such an extent that there has not been, until recently, a contemplation of the negative economic effect on the areas that had been served by the previous innovation in transportation, the canals. Both of these errors may be taken as evidence of the power of symbolically important events to dominate thinking about an era, accruing both

[19] Mazlish, *Railroad and the Space Program*, pp. 34–35.
[20] *Ibid.*, p. 22.

credit and blame without necessarily having played any direct role.

The most exciting and lasting second-order consequence of the railroads may well be the unintended consequences of the economics and management of such a large and new enterprise. The sums of money required for the development of the railroads were so large that they could not be raised by the then established means. As a result, the railroads stimulated the development of a capital market. This institution then took on a life of its own, and became a key instrument in the development of other areas of the economy.

The efforts of the railroad management to coordinate the activities of so vast an enterprise spread over such a wide area brought about the modern organizational forms whereby large enterprises are run today.

Operating hundreds of miles of line, with, for the time, enormous capital investments . . . the railroads had to invent totally new administrative devices. . . . For example, the large distances and complicated traffic management created problems of safety and scheduling. To solve these problems, a strict administrative hierarchy, run by professional administrators (who, at first, were engineers), was necessary. . . . Chandler and Salisbury . . . trace the development of a decentralized administration on the Pennsylvania Railroad, with its division of line and staff duties, so admirably suited for the great distances on American railroads. Such a development, however, was not inevitable, as the New York Central's growth as a centralized bureaucracy, more on the English short-distance model, so well demonstrates. . . .

Centralized, or decentralized, managed by financiers or engineers, the American railroads created the modern form of administration, wherein business activity moved away "from organizations run by entrepreneurs with the aid of personal trustees, relatives, and the like to corporations with a systematized bureaucratic management." Further, by contributing forcefully to the creation of a securities market, the railroads further pioneered in the separation of ownership and management which has become so characteristic of the American corporation. . . . the new forms of professional . . . management and labor. With the necessity of administrative expertise, the Horatio Alger rise from round-house to corporation president was no longer a meaningful reality.[21]

As one contemplates the nature of these effects, he sees that the value of historical analogies lies not only in the similarities they may reveal, but in the extent to which one is prompted to explore systematically the meaningfulness of differences. One can say that the lesson to be learned is that new sociotechnical inventions of a

[21] *Ibid.*, pp. 24–25.

large scale will pose new management problems and therefore may stimulate new organizational forms and management practices. But one cannot specify what the form of the new organization will be, or the nature of the new practices.

One would also suspect that the new organizational forms and management practices will place demands of some sort on the environment and bring about changes in broader social practices. For example, in order to make the administrative form of the railroad practical, it was necessary to establish standard time zones throughout the country and to provide rapid communication via telegraph. Standard time zones might not have occurred for a long while if it had not been for the needs of the railroad, and the telegraph would have spread more slowly through the country.

While we do not treat the matter specifically in this volume, it is already quite clear that the tasks and work forms of the aerospace industry are indeed producing new organizational and management forms and procedures. For example, the "project management" system involves the creation of *ad hoc* organizations within stable organizations, with the project organization in a relatively constant state of change in adjustment to the evolving demands of the project.[22]

The space program does have "stages," as Mazlish suggests, yet they are not so clearly temporal. In a sense, the space effort is two sets of activities, the first being a massive socio-technico-economic effort to produce space hardware (and train spacemen), and the second the activities in space itself, some practical, some political, some military, some scientific, and some adventurous, in varying mixtures. Clearly these two sets of activities overlap in time, but the relative importance of their impact is indeed time phased. As I will discuss in Part II, which is concerned with the actual effects of the space program, the presently traceable effects have to do mainly with the activities here on earth, or with the anticipation of activities in space. (This is not to deny such real, concrete effects as better weather forecasting, transoceanic telecasting, and so on.)

Furthermore, the matter of "national style" is of importance, particularly if one is to extend the concept to include the style of national leaders. Generally speaking, writers concerned with the social impact of technology speak of a one-way causation, forgetting the extent to which any technological effort itself is a social

[22] For one early discussion of this development see Fremont E. Kast and James E. Rosenzweig, *Management in the Space Age* (New York: Exposition University Press, 1962).

product. The contrasting styles and emphases of the Soviet and American space programs have been commented on endlessly. In a similar fashion, development of the railroads in England and in the United States was carried out in markedly contrasting fashions, reflecting the differences in resources, problems, and probably the style of the people. Project Apollo, or perhaps more broadly speaking, the use of *some* dramatic goal as a way of symbolizing American efforts, was unquestionably a matter of the style of John F. Kennedy in his effort to "get the country moving."

But probably the most important over-all lesson learned from the investigation into historical analogy by Mazlish and his associates was the nature of the process. The use of analogy is not a matter of "learning from history," but rather a transactional process in which past, present, and future illuminate each other. The searcher for analogy turns to the past with questions growing out of the present and out of those futures which he can foresee. In the process he sees the past in a way in which he had not seen it before. But the very effort to pose the questions to be asked of the past forces one into a clearer conceptualization of what he thinks he may be concerned with in the present and the future, and to a more precise definition of terms. In Mazlish's case, he was prompted to define the space effort as a "social invention" rather than "exploration" or "discovery." The purpose of dealing with the problem of the use of analogy, particularly historical analogy, was to illustrate the task of anticipating possible future states that might flow from present action. Anticipation, however, is but one aspect of the process of planning and control. If we take seriously our statements about the necessity of feedback for the rapid detection of error, then we must also take seriously the establishment of measures to effect this detection. It is with this second aspect of the process of planning and control in mind that we turn to the matter of measuring social change over time.

3

Measurement Over Time

The fact is that neither the President nor the Congress nor the public has the kind of broad-scale information and analysis needed to adequately assess our progress toward achievement of our national social aspirations.[1]

If we are to consider the contributions of the space program to our national goals, and we have no "broad-scale information and analysis needed to adequately assess our progress toward achievement of our national social aspirations," then how are we to know whether we have done ourselves good or ill? How can we measure effects and exercise control over the future without basic data? Senator Walter Mondale's speech quoted above was made approximately four years after the American Academy's Committee on Space began work on the volume, *Social Indicators*.[2] Our work was motivated by the same consideration that prompted the introduction of Senator Mondale's bill: the realization that amid all the talk about the probable consequences of our program of space exploration, there were precious few yardsticks to measure whether such consequences had actually occurred.

I have mentioned the extent to which new approaches to planning and control have evolved rapidly over the course of the past few years while our own work has been in progress. The same is true of interest in the problem of measuring social change, and of social measurement in general. The manifestations are many.

The National Commission on Technology, Automation, and Economic Progress pointed out that our ability to measure social change has lagged behind our ability to measure strictly economic change. This Commission called for better measures in four areas:

[1] Senator Walter Mondale, in his speech introducing "The Full Opportunity and Social Accounting Act of 1967," *The Congressional Record, 113*, No. 17 (February 6, 1967).

[2] Raymond A. Bauer, ed., *Social Indicators* (Cambridge, Mass.: The M.I.T. Press, 1966).

(1) social costs and net returns from innovations; (2) social ills, such as crime and family disruption; (3) a "performance budget" of defined social needs, such as housing and education; and (4) economic opportunity and social mobility.[3]

The United Nations expressed its concern for cross-national comparisons in a U.N. report on *Methods of Determining Social Cost Allocations*, calling for "the development of a comprehensive set of criteria that will take account of both economic and social considerations, not by forcing the one kind into the mould of the other, but by integrating them at a higher level of abstraction."[4]

The Russell Sage Foundation has inaugurated a program to review our measures of social change. And the *Annals of the American Academy of Political and Social Science* has two issues on social indicators in progress under the editorship of Bertram Gross.

Most important, President Johnson, in his 1966 message on health and education, gave the following instruction to the Secretary of Health, Education, and Welfare:

> To improve our ability to chart our progress, I have asked the Secretary to establish within his office the resources to develop the necessary social statistics and indicators to supplement those prepared by the Bureau of Labor Statistics and the Council of Economic Advisors. With these yardsticks, we can better measure the distance we have come and plan for the way ahead.[5]

In mid-1966, a Social Indicators Panel was established in HEW to carry out this mandate.

At last there has been an awakening to a state of affairs that has concerned statisticians and social scientists for decades. The reasons for this awakening are undoubtedly many. Possibly the most concrete of these reasons is the spread of the "P.P.B.," the McNamara sponsored system of planning, programming, and budgeting, which originated in the Pentagon and spread throughout the government during the years 1966–1967. But this came in the wake of our successful decades of experience with the system of Economic Indicators, which had been used to guide the U.S. Economy in the years after World War II.

---

[3] Report of National Commission on Technology, Automation, and Economic Progress (Washington, D. C.: U.S. Government Printing Office, January 1966).

[4] Report of the Secretary-General to the Sixteenth Session of the Social Commission, Economic and Social Council, *Methods of Determining Social Cost Allocations* (March 31, 1965), p. 10.

[5] President Lyndon B. Johnson, Message to the Congress on domestic health and education, May 1, 1966.

The spread of P.P.B. came on the heels of lively exploration of the limitations of the system of Economic Indicators. The combination appears to have created a concern for the possibility of a "New Philistinism" — a term Bertram Gross has used to indicate the dangers of measuring cost and benefits of government programs in strictly economic terms. Furthermore, in proposing the program of the Great Society, President Johnson placed an emphasis on the quality of life. All of these circumstances created an atmosphere in which suggestions for improved measurements of social phenomena were listened to. Daniel Bell, who has long had a concern over the fallability of our measures of social change, was a member of the Commission on Technology, Automation, and Economic Progress and pressed for the incorporation of the plea for better measurement in the report of that body. Our own work, which we began with modest aspirations, was welcomed to an extent far exceeding our hopes.

The Task

As we saw the issue, the problems of anticipation, feedback, measurement of effects, and evaluation of the nature of changes in the society that might result from the space program are closely interrelated. If a change that has been anticipated takes place, it must be manifested in something that can be observed. If the space program is responsible for that change, there must be some way of establishing the linkage. And, there must be some way of evaluating the change in terms of the society's values and goals, and its impact on other programs. Here again I cite at length from our view of the problem early in 1963:

Considering the confidence with which social critics of this and past eras have commented on trends in society, one might assume that the basic task of being able to measure "good" and "bad" trends in society is well in hand. In point of fact, this problem has scarcely been approached. The field of economics is an exception against which other areas may be evaluated. Economists have, over a number of decades, developed a series of indicators which are used to evaluate and anticipate trends in the development of the economy. Economists differ in their evaluation of the meaning and implication of such indicators. However, relatively speaking, there is consensus on *what* should be observed and how it should be observed. *Some* isolated trend series have been developed in various other areas of society, notably in fields such as public health. To discuss the technical difficulties of most of these series, e.g., the incomparability of data from various time periods, might seem captious, since what is initially important is the lack of consensus as to what should be measured. This lack arises in part from a deeper lack of consensus—lack of consensus on goals, purposes, and the nature of society.

Probably some reasonable agreement could be reached on some range of goals and values on which there is consensus. Let us at least assume this. A new series of problems develops immediately. Many goals and values are of such a level of abstraction that they give us little guidance as to what to observe. Take as examples "national unity," or "the development of the fullest potential of the American people." These are values with which few Amricans would disagree. Yet there are wide varieties of events which one might observe and which bear on these values, and they would not all be moving in the same direction at the same time. Consequently, the resultant observations would be hard to coordinate.

Even the strictly technical problems are formidable. An economic index such as cost of living must be constantly readjusted to account for the changing composition of the standard market basket. In a similar manner, even so simple a phenomenon as change of educational level of the population cannot be evaluated properly without taking into account differences in the content of the curriculum, the uses to which education is put, and the meaning of extending education to successively less intelligent segments of the population. The difficulties of developing indices of social pathologies for, e.g., crime, mental illness, mental deficiency, divorce, and suicide, are enormous because of the shifting of definitions of such categories.

It might be assumed that when there is agreement on our goals, and values are agreed upon, we can decide what observable events will be "surrogates" for these goals and values, and we have solved the technical problems of measuring and comparing them over time, that our job is done. But, this is not so. If we have solved all the foregoing problems we will then be in a position, hopefully, to evaluate the trend of events *up to the present time*. But policies are set for the future. Put most simply, if an observation is made today and assessed as to what it means [for] tomorrow, this assessment must be made on the basis of the flow of events between today and tomorrow—and on some assumption as to what tomorrow and the intervening time will be like. Thus, an increase in desire for technical education among the young may be regarded as a "good thing," but we can be confident of its desirability only to the extent that we are confident that the need for such skills will be important at the appropriate time in the future.

Viewed in this way, social indicators are simultaneously devices for prediction, and also yardsticks of progress which can be used only on the basis of assumptions about the future. (It seems to us impossible to conceive of any concrete situation, in terms of the total nature of society at present no matter how desirable, that could not prove undesirable under some conceivable future development of society.)[6]

The foregoing passage reflects our early concern with the prevailing modes of evaluating social trends, this concern being based both upon the absence of trend data on many aspects of our society that are of interest, and upon the difficulties of making proper inferences even from good data. As a result we asked ourselves what sort of societal information system would be necessary for making such evaluations of the social impact of space exploration as were involved in the various commentaries.

The questions we asked ourselves were of this order: What are the limitations to the existing measures of social change? What sort of societal information system might represent some sort of "ideal"? Given some ideal of a broad societal information system, what would this mean to an agency such as NASA with its peculiar interests and responsibilities?

Our aspirations seem to have been less than was possible; we underestimated what could be done. My personal objective in organizing the work of my collaborators was that we should produce a critique of the state of social statistics and of the inferences drawn about social progress or social retrogression sufficient to instill a greater sense of responsibility into the evaluations of programs such as the space program and greater humility as to our ability to identify and/or anticipate their effects. Our efforts to delineate a societal information system were presented in the following spirit: "If people want to make the sorts of statements they make, and want to back them up with a reasonable amount of evidence, here is the way to do it. This is not to say that we expect NASA or any other agency of the U.S. government to take us literally, but they may want to borrow a suggestion here or there." As I have already hinted, and will spell out in more detail later, the tides were running with us, and proposals such as we made are presently being taken more seriously than anyone had reason to expect in 1963.

Before plunging into a consideration of the present state of social statistics, I should outline briefly what the basic components of a societal information system would look like so that the succeeding discussion can be kept in perspective. If any agency such as NASA

---

[6] "Space Efforts and Society: A Statement of Mission and Work," a document of The Committee on Space Efforts and Society of the American Academy of Arts and Sciences (Boston, January 1963), pp. 15–16; see the Appendix, pp. 206–208.

really felt that it had a mandate to increase the beneficial consequences and dampen the less desirable ones, what sort of an information system could it possibly contemplate?

First, there should be a series of "social indicators," i.e., an ongoing statistical series that would make it possible to chart the continuing trends in major features of the society, to judge what is happening to the performance features of the society, and to judge its future capacity to perform. However, not all relevant events would fall into such series. Some one-time or seldom occurring events are important; therefore there should be a provision for stand-by research facilities to make before and after measurements, if possible, of the effect of such events as a Presidential assassination or the launching of the first communications satellite. As is generally known, our facilities for making such studies are even poorer than our social trend statistics. The Center for Disaster Studies at Ohio University is the only group working systematically on this problem.

We may regard the regular statistical series, and the study of rare events of individual importance, as our primary information system. This in turn should be complemented by provision for analytic studies that relate the long-term trends in society to events contiguous in time and/or space and that attribute a causal relationship among them. If delinquency increased during a period of urbanization, then it would be assumed that urbanization increases juvenile delinquency. And, finally, there should be some sensible provisions for the use of information. There is a prevailing notion that all information is good, but all information in all forms at all times is not equally useful. We must give some thought to getting that information which is useful, getting it in an appropriate form, and reporting it to the right place and person.

These, then, are the components of a societal information system: basic social trend data (including, of course, economic data); stand-by facilities for researching rare and difficult-to-anticipate events (this would include program evaluation to the extent that we were interested in the consequences of programs as compared to assessing them); regular facilities for analytical studies of the process of change; and the engineering of sensible information systems for each of the using parties, whether they be policy makers, persons with operations responsibilities, or social critics.

While I have enumerated these four elements of an informational system, their successful functioning is dependent on still another, and broader, consideration, namely the model of our social system that we should employ. This will determine selection of the variables to be measured, the evaluation we put on the present state of affairs, and the judgments we make so as to act rationally to produce future states of affairs that are desirable. A model of the economic system is at the core of the successful use of the economic indicators. But this is not yet available for the social system. Whatever reservations we may have about the economic model, and whatever claims sociologists may make for their models of the American social system, the fact still remains that there is not sufficient consensus to warrant the adoption at this point of a single model. It is to be hoped that an adequate set of social indicators will provide the basic societal data permitting such a model to be developed.

The State of Social Statistics

Forsaking for the moment consideration of a total societal information system, I will concentrate at this point on the series of ongoing historical records of economic, social, and political events in our society. The following statement of their limitations is in no sense a general criticism of the people and organizations who currently assemble, process, and interpret our social statistics (except for those social critics outside the statistical establishment who use such data in a fashion that the originators of the data would not sanction). Social statisticians in general have been the first persons to stress the limitations of the data they generate, and they are pushing hard to improve their trend data.

1. *Many phenomena we apparently regard with seriousness simply are not represented in our trend statistics, and this "non-representation" increases as we move into areas of more recent concern, such as science, technology, and what we may call the "quality of life."* Albert Biderman compared the statements of 81 national goals formulated in 1960 by the President's Commission on National Goals with our two main sources of trend statistics, *The Statistical Abstracts of the United States, 1962,* and *Historical Statistics of the United States: Colonial Times to 1957.*

The loosest criterion of relevance was applied; the only objective was to determine whether the indicators in these volumes were related to phenomena that were at all pertinent to each goal statement. The indicators did not need to be in any sense an index of attainment of the specific goal.

For only 59 percent of the goal statements are any indicator data in these sources judged pertinent. . . . Interestingly, the two domestic goal areas probably most directly related to space activity—"arts and sciences" and "technological change"—are among those for which these sources provide the least indicator coverage.[7]

Certain goal areas are rather well covered. Of five educational goals mentioned by the President's Commission, all five are covered. Nine goals pertaining to economic growth are all represented in the existing statistics series, as are ten health goals and four out of five goals in the area of agriculture. The technical adequacy of such measures will be discussed below. However, at least some areas of traditional statistical concern are well represented.

As Biderman indicated, arts and sciences and technology are poorly represented. A total of 13 goals are suggested in these two categories, and only three are to be found in the trend data offered. Also poorly represented are goals pertaining to the democratic process, the democratic aspects of the economy, the status of the individual, and living conditions.

In other words, our social trend data are conspicuously weak in those areas of our society about which we have become more recently concerned. In effect we have only the very weakest bases for making trend statements about what is happening to many aspects of our life.

2. *Many of the trend data are at best approximations of what we are interested in. They range in their inadequacy from being weak indicators of what they purport to measure to being actually or potentially misleading.* The major source of the inadequacies of present statistical series stems from the fact that most of these series — the census being a notable exception — are based on data gathered routinely in the course of administering the affairs of the society, rather than by direct measures of the phenomena about which we are concerned.

Even economic series, which are in some ways our model of what can be done, suffer from these technical inadequacies. At the heart of our economic statistics is the concept of Gross National Product. This concept, the GNP, is based on the measurable economic transactions that get recorded in our economy. Reflecting the technical problems of this measure is the standing joke among statisticians that a man can lower the GNP by marrying his housekeeper, thus converting her from a gainfully employed person

[7] Albert D. Biderman, "Social Indicators and Goals," in Bauer, *Social Indicators*, p. 87.

to someone who does the same work without a recorded economic transaction. Similarly, homemade clothes, house repairs done by the occupants, and so on, do not enter into the Gross National Product even though they represent economically productive activity. At present, efforts are afoot to estimate the worth of the unpaid work done by housewives for potential inclusion in our estimate of the productivity of our and other economies. It would appear that the inclusion of estimates of the value of this work will immediately produce a jump of tens of billions of dollars in the apparent GNP.

I do not feel personally competent to comment on the extent to which such technical difficulties have an impact on our conduct of the economy, though it is obvious that professional economists are concerned.[8]

With respect to other series, however, the layman can quickly grasp their difficulties. For example, until a few years ago unemployment was measured by the number of persons receiving unemployment compensation. When a person exhausted his unemployment benefits, he automatically left the rolls of those recorded as unemployed. The minimal difficulty of this measure was that it failed to include those persons who were most truly unemployed. Maximally one could imagine a situation in which unemployment in a distressed area had been actually increasing while it was apparently decreasing because only the relatively recently unemployed were being recorded. In recent years this deficiency has been eliminated by the introduction of a regularly conducted sample survey of the employed and unemployed, thereby freeing our estimates of these key parameters of our society from the limitations of administratively gathered data.

One of the most publicized of our measures of social change, the Index of Serious Crimes, is based on administratively gathered statistics — the crimes *reported* by the police. Unreported crimes, or crimes reported by other sources, are not included. Further distorting the picture is the current system of classifying reported crimes, which tends to create the image of a society in which violence is increasing at a rapid rate. Crimes of violence, constituting a very small portion of the Index, are unfortunately lumped together with automobile thefts and larceny, which accounts for most of the apparent increase in violence in the Index. However, it

---

[8] Cf. the discussion of the concerns of economists in Bertram M. Gross, "The State of the Nation: Social Systems Accounting," in Bauer, *Social Indicators*, pp. 165 ff.

should be noted that the recently submitted report of the President's Commission advocated a look into better measures of criminal activity in our country.

Deficiencies in the measurement of education and health have also been recognized, and steps are under way to improve them. In education not only are measures technically inadequate, they also perpetuate the confusion between costs and investment. As we plot the educational levels of our people over time, or the comparative education of various groups in the population, we view the data from two perspectives: as an index of the benefits we confer on our people, and as a measure of the capacities of these people, to share and produce culture, to perform productive work, and so on.

From the first point of view, benefits conferred, we are becoming aware of the imperfections of this measure. Eight years of education in a segregated school is not the equivalent of the same term in an integrated or all-white school. Further, as a measure of the capacities of the population for productive work, number of years of schooling achieved is an exceedingly poor measure of a man's present capacity to perform the tasks of the work force at a given period of time. It is probably an even worse measure of his capacity to learn new knowledge and skills.

Under Secretary of Health, Education, and Welfare, Wilbur J. Cohen, has had this to say:

When we survey the voluminous, yet unsuitable, data now available for assessing the products of our education, we must conclude that practically none of it measures the output of our educational system in terms that really matter (that is, in terms of what students have learned). Amazement at this revelation of the tremendous lack of suitable indicators is almost overshadowed by the incredible fact that the Nation has, year after year, been spending billions of state and local dollars on an enterprise without knowing how effective the expenditures are. . . .[9]

There is an established technology for making direct measures of knowledge, ability, and achievement. While there has been controversy over the use of tests devised by psychologists, the fact is that they are in no way inferior to the criteria whereby students are promoted in and graduate from the schools they attend. Properly used, they can be much more adequate measures of the present status of the population's skills, knowledge, and so on. Such

[9] Wilbur J. Cohen, "Education and Learning," *The Annals*, ed. Bertram Gross (Philadelphia: American Academy of Political and Social Science, 1967), pp. 79–101.

measures, gathered on a sample basis can give us a far more precise estimate of the capacities of the U.S. population on relevant dimensions than we can gauge by the record of their past educational history.

At this point in time, a considerable number of studies have conducted various measures of ability on national samples, usually of children. The Carnegie Corporation study of quality of education being conducted by Ralph Tyler is the most current example. Such measures are entirely feasible and, since they must be conducted on a sample basis, are not excessively expensive once the procedures are established. Most relevant, they focus our attention on the fact that we are interested in our citizens as a national resource as well as beneficiaries. More appropriate conceptualization of what we are interested in, and proper methods of measurement, will enable us to make more sensible distinctions between the proportion of health, education, and such expenses that should be regarded as cost, and that which should be viewed as a capital investment.

A plea for "better conceptualization," such as I have suggested, almost certainly has an unearthly, abstract, and academic ring. Happily I can report that this is exactly what has been going on successfully in the National Center for Health Statistics of HEW.[10] Until the mid-1950's, our primary concern was with our ability to keep people alive at various ages; *mortality* was the relevant criterion of health. Since then, considerable attention has been devoted to *morbidity* or departures from health of a nonfatal sort. But unhealth is not as clear-cut a proposition as death. It may be thought of as some degree of organic or psychic malfunction susceptible to medical diagnosis. It may also be defined in terms of the individual's incapacity to perform with some degree of effectiveness in his various roles. Several of the concepts of disability or unhealth employed in the National Health Interview Survey, conducted monthly, have related to the ability of the individual to perform his roles—as a worker, housewife, student, mother, father, and so on.

The immediate physical condition of a person is a matter of concern for his own state of comfort, may have economic consequences because of his need for care and medication, may present a threat to others if his condition is infectious, may serve as in indication of our ability to conquer certain diseases, and so on. However, if we

[10] See Daniel F. Sullivan, *Conceptual Problems in Developing an Index of Health* (Washington, D. C.: National Center for Health Statistics, Series 2, Nov. 17, 1966).

are concerned directly with his ability to contribute to the society (or conversely with the effect of sickness in general — a given disease or a specific epidemic — on economic production or on the conduct of the day-to-day business of the society), then a measure of the extent to which people are able to perform their roles is the most relevant criterion of health we can have. At a certain stage of its development, cancer is less likely to restrict role performance than is the common cold.

The illustrations offered are intended to substantiate two points. For many aspects of the society there simply are no comparable measures over time. This is essentially true of those aspects of the society about which we have developed increasing concern in the Space Age, namely, science, technology, and various concepts of the quality of our personal lives. In addition, many of the measures we are accustomed to using are in varying degrees only approximations of what we are interested in, and we can in fact make more direct and adequate estimates of them. Again, the inadequacies are selectively biased against the noneconomic aspects of life, for instance, the costs of education are easier to document than are the personal benefits to the individual.

Assignment of Values

What should be measured? How should it be measured? Neither question could or should be answered neatly at this point in time. All we can do is suggest the sort of criteria to be invoked and the direction in which things are likely to go.

The data we want obviously are dependent on the goals we set for ourselves, on the type of people and the type of society we are. A society primarily bent on achieving military power would be interested in measuring different things from one bent on maximizing religiosity, aesthetic experience, or material comfort. Therefore, we need to know the values of the members of our society, the programs or "national goal" that *ought* to represent attempts to serve those values, and the interrelationship of the programs and values.

The knowledge of people's values and aspirations is a key to planning and control. It tells us on the one hand what our people want, and on the other the sorts of programs they will support and/or tolerate. The study of values and aspirations can be technically difficult and politically sensitive. Yet, as difficult and sensitive as the study is, it continues to be done, no matter how inadequately. Considerable sophistication has developed for the interpretation of such data. For example, politicians know that opinion-poll data showing a relatively low level of public support for the space program would not be an urgent mandate for dismantling NASA. At no point will knowledge of such values and aspirations, of themselves, be a clear guide to a specific course of action. Yet better knowledge of people's values can give us an improved basis for judgment on many issues. As for aspirations, it wasn't until *after* the Watts riots in California that it was generally realized that the aspirations and expectations of Negroes had been raised beyond our capacity to meet them. Whether or not studies of values, expectations, and aspirations should *presently* be made by a Federal agency is, however, another matter.

In the market place, the conventional and probably proper way of putting a value on goods sold in the market is the amount that people will pay for them. It is characteristic, however, of many public investments that their products (for instance, weather forecasting) can be shared by an untold number of people; thereby the "value" of such goods and services is a function of the number of people who will benefit by them. But, in the absence of a direct measure of value in terms of the utility of public goods and services to those who use them, the prevailing practice has been to evaluate a public investment solely in terms of its cost. Economists agree that utility cannot be measured solely in monetary terms (at a minimum, a dollar is worth less to some people than to others); so do the disciples of P.P.B. Hence, to the extent that we can evolve measures of utility that are a direct reflection of the benefits people perceive themselves as receiving, the more adequate the basis we will have for evaluating public investments whose value is dependent upon the number of persons benefitting.

Again, we are not proposing the tilling of entirely unplowed ground. Present methods of program evaluation often involve research on the benefits conveyed. The poverty program, for example, is presently the object of such evaluations. The point I want to make is that as we develop increasingly more adequate measures of people's values and expectations, we will also be able to develop more appropriate measures of benefits. (We would judge differently the effect of a physical fitness program on a boy who wants to become an engineer from its effect on a boy who wants to become a professional athlete.) The combination will enable us to plan better and evaluate better.

Cost, or Investment in the Future

Another issue on which more appropriate measures will aid planning and evaluation is that of the extent to which an expenditure should be regarded as a cost or as an *investment*. For example, by and large expenditures on education have been regarded as a public *cost* producing a private benefit. However, a business firm that builds a plant will regard this as a capital *investment*, which increases its assets. Corporate accounting draws a reasonable distinction between money consumed in the form of current costs and money invested to yield some future return. But should an investment in plant equipment be treated differently from an investment in training? The idea of federal capital budgeting has been around for some time. It would make more sense if we had a better basis on which to decide between an investment and an expenditure.

The first step in reaching wise decisions as to the extent to which social expenditures should be regarded as costs or investments is, prosaically enough, to recognize the issue. For many decades, there was a prevailing disposition to regard all expenditures outside of the market segment of the economy as costs, simply as a matter of discipline against "extravagant government spending." The most stereotyped of this sort of thinking is happily on the wane, and the issue can now be broached with some possibility of generally constructive thinking on the matter.

The second step is that which we have been considering, namely, the better conceptualization of the various phenomena on which public moneys are spent, and the fashioning of measures that reflect this conceptualization.

The third and fourth steps are to establish a more formal set of "accounts" with which to keep track of the results of public expenditures and programs, and beyond that to develop a social system model parallel to that of the economic model as a basis for better evaluating the future consequences of these expenditures and programs. The last of these is the ultimate refinement of the series of steps, and I will return to it in the last portion of this chapter. In the meantime we may turn our attention to the third step—a system of social accounting.

Social Systems Accounting

In the treatment of a system of social accounts, I will touch only lightly on a topic dealt with at some length by Bertram Gross in

"The State of the Nation: Social Systems Accounting"[11] and by Senator Mondale in "The Full Opportunity and Social Accounting Act of 1967."

Gross proposes a system for social accounting that would fall into two broad categories, system structure and system performance. For purposes of exposition I will treat them in inverse order. System performance would include measures of the extent to which, at a given point in time, we have been able to meet the values and aspirations of our citizens. The other broad category, "system structure," would include measures of the society's capacity to perform its functions in the future.

Measures of "system performance" would be our present measures of economic performance, measures of the distribution of economic benefits throughout the society, the education and economic opportunities we have afforded our people, the environment we have created for them to live in, the cultural opportunities we have offered, the extent to which we have created a peaceful world, the level of health of the nation, and so on. Any state of affairs that people value and aspire to is a potential candidate for inclusion in this list, as is any state of affairs they would like to avoid.

The following passage from Senator Mondale's speech illustrates some of the "performance" data we need to evaluate how well we are serving the needs of our citizenry.

How many Americans suffer in the squalor of inadequate housing? How many children do not receive educations commensurate with their abilities? To how many citizens is equality of justice denied? How many convicts in our penal institutions are barred from rehabilitation that would allow them the opportunity to re-enter the mainstream of life? How many physically handicapped and mentally retarded are unable to get training to achieve their potential? How many individual Americans are denied adequate health care? How many are breathing polluted air? These are some of the possible indicators that might be considered in the social accounting.[12]

One of the standard criticisms of the welfare state has been that it creams off present benefits "at the cost of future generations." This is, of course, a reasonable issue to raise. For this reason, Gross would parallel measures of future capacity to measures of performance. Such measures would include conventional measures of the strength of the economy, and less usual measures of the state

[11] Gross, in Bauer, *Social Indicators*, pp. 165 ff.

[12] Senator Walter Mondale, "The Full Opportunity and Social Accounting Act of 1967," *The Congressional Record, 113,* No. 17, (February 6, 1967).

of science and technology, of the state of skills and knowledge in the population, of its health, of the extent to which our institutions make it possible for all members of the society to make their potential contribution, and so on.

It is clear that in a scheme such as this, many of the items we presently regard in a unitary fashion would have entries on both sides of the ledger. The education of our people represents in some ways a cost, a measure of performance, and in some ways a benefit, a measure of future performance capacity. The same is true of investments in health. Or, an improvement in the stability of the families of underprivileged groups, such as has been advocated in recent times, can be seen as contributing present benefits to those groups, and future benefits to the productivity of the society as a whole. It is here, in providing for appropriate entries on both sides of the ledger, that better conceptualization and measurement become crucial.

But, again I return to a familiar theme. These entries in the system of social accounts will not interpret themselves. An accounting scheme such as Gross and Senator Mondale propose is a tabulation of the rise and fall of "good things" and "bad things." The evaluation one puts on them is a matter of the over-all social system model he uses, whether explicitly or implicitly. It will also depend on the availability of information that the present series of trend data themselves do not provide: information about important events, the effects of public programs, and the causal relations among such elements.

Ad Hoc and Stand-By Research

Biderman, in the second of his essays in *Social Indicators*, makes a plea for "Anticipatory Studies and Stand-By Research Capabilities."[13] We may lump his concern together with the need for better program evaluation and for analytic studies that link programs such as that of NASA to the social change that would be recorded in a system of social indicators.

Biderman's main concern was for the fact that many important events, e.g., the assassination of a President, changes in draft policies, tax increases or decreases, and the discharge of large numbers of men from the armed services, would not fall in any established statistical series. Yet, such events can be anticipated and

[13] Albert D. Biderman, "Anticipatory Studies and Stand-By Research Capabilities," in Bauer, *Social Indicators*, pp. 272 ff.

their effects measured. It may be said that our facilities for such measurements are very poor indeed. But so are our practices in assessing the effects of government programs, and for making analytic studies that establish the actual linkage between programs such as space exploration and the events that reputedly flow from them.

The idea of developing a broader, more comprehensive set of social statistics has caught on. In view of this it seems more appropriate for me at this point to try to put this movement into perspective than to argue for that which seems generally accepted.

There is little argument to be made against better information *per se*. There is also little argument but that the basic case for social accounting has been recognized for a long time in this country — since the beginnings of the Republic.

However, the technology for a "great leap forward" is relatively new; it is the sample survey.[14] Most of the new data, which is being gathered and which should be gathered, is better garnered from people than from administrative records. Good data on unemployment had to be gathered from a cross-sectional sample of the population. Data on people's performance of their social role as a consequence of sickness could only be gotten from the people themselves. Direct measures of knowledge, skill, and ability must be derived from people in whom these attributes are lodged.

But, at the same time that interest in social indicators has been growing, interest of a parallel nature has developed in "data banks." While the two interests have a good deal in common, there are some vital differences between them. The two movements are in agreement with respect to the goal of making information available for decision making. The data bank movement, however, focuses more on the computer and its capacity to store and retrieve enormous amounts of information. By and large, the data bank movement has been relatively more concerned with the question of how to get data in and out of the computer, while the social indicators movement has concentrated on the proper relationship of data to its use, with a heavy emphasis on having new data series be more appropriate to the purposes they are put to. The proper label for the social indicators movement is more that of information system than of data bank.

It is obvious that the computer will be an important tool no matter which of these orientations is involved. And both lines of

[14] Raymond A. Bauer, "Social Indicators and Sample Surveys," *The Public Opinion Quarterly*, 30 (Fall 1966), 339 ff.

activity, the storage and retrieval of available data and the gathering of new, more useful data, will complement each other.

However, the prospect of the storage and retrieval of vast amounts of data accumulated on each individual in the country has evoked considerable and understandable anxiety over the protection of the privacy of the individual. There is good reason to believe that the public officials responsible for such data banks — at least on the Federal level — are sufficiently vigilant that they can duplicate the splendid history of the Bureau of the Census in protecting the privacy of the individual. That issue aside, however, I should like to point out that the types of additional information that are being suggested by the advocates of a comprehensive system of social indicators do not add to, and would probably subtract from, this anxiety over the existence of "thick dossiers on everybody." Data gathered by sample surveys involve rather small numbers of people, usually interviewed on a one-shot basis. There is no need to retain any means of identifying them at any future time. It does seem that certain types of data, such as those pertaining to life opportunities of individuals for education and employment over many years, demand that people be studied on a longitudinal basis. However, the numbers of such people — while they might constitute a very large sample of several hundred thousand — are still far fewer than the "everybody" about whom the opponents of dossiers are concerned.

On the whole, it can be argued that the movement for better social indicators can be used to alleviate anxiety over the invasion of privacy. This does not follow automatically, but is dependent upon the extent to which one can communicate to the public that what is at stake is not a vast information file about individuals but measures of key parameters of the society in which those individuals live, based on information that only those individuals can give. In a word, the individual is being asked to "inform" on society, but should be protected against informing on himself.

Another issue that arises immediately as one talks about added data series is that of the burden they will place on those individuals asked to supply the data. I can only reply that samples are usually such a small proportion of the total population that one could gather quite a few data series on a quarterly or yearly basis without any one individual being excessively burdened. The reality of the problem should not be swept under the rug, and it is one on which men will have to exercise some common sense.

Cost is not too serious a problem initially. The Federal statistical budget is in the vicinity of $160,000,000. Doubling or tripling this figure would provide for the addition of many new statistical series, and the total cost would still be small relative to the value of such information. The issue of cost would become pertinent as the data proved their value. Surveys, such as the monthly survey of health carried out under HEW, are based on samples of a size adequate to permit inferences about the state of the U.S. population. However, as their results become known and valued, smaller governmental units, such as the individual states, request finer breakdowns of the result than present sizes permit. This is known in the statistical trade as the problem of "disaggregation." If the data prove so useful that larger samples are called for, then the problem of balancing the utility against the cost will have to be faced in individual instances and decisions made on the merits of the individual cases.

More serious than any of the above considerations is the premature launching of data series without sufficient experimentation with alternative measures. There are many persons skeptical of our ability to measure such "intangibles" as people's values and expectations, or some of the other variables that have been proposed. Unhappily it is not the skeptics who will try to make such measures but the enthusiasts. The difficulty with concepts such as these is not that they are "intangible." All concepts are literally intangible. The real problem is that they are complex and are subject to great latitude of definition as well as many potential types of observations, which may stand as surrogates for such concepts. The concepts and measures and the inferences that can be drawn from them can be hardened only in use, and the starting of any data series incorporating such variables should be accompanied by a wide program of experimentation on alternative approaches. The danger is that enthusiasts for one particular measure will not be sufficiently concerned with the need for this sort of experimentation.

In closing this chapter, I would like to return to the question of a model of our social system. I have said several times that, lacking any consensus on such a model — and I am certain we are far from such a consensus at this point — we can establish data series only for variables about which there is generalized agreement. It is to be hoped, however, that the existence of more and better data over a period of a few decades will permit the development of a model as generally acceptable as that with which economists have worked for the past decade or so.

I would hesitate to undersell our present state of knowledge of social systems, or of models of smaller segments of the society such as the family and the firm. There are many such models and theories which can be used to identify important societal variables and which in turn can be used to interpret such data. What we have to look forward to is some decades of choosing among and refining of these ideas and the emergence of some one workable model that can be agreed upon — only again to be revised.

In this first section of the book we have considered the over-all question of second-order consequences of technological innovation from a highly general point of view, spelling out what we saw as relevant for thinking of this as a problem of planning and control and suggesting the aids one might seek in anticipating and measuring such consequences. This can be considered as no more than a prefatory exercise for a look at actual consequences of the space program. We hope it was a worthwhile exercise and undertook it because we were not aware of any other comparable treatment. Now, however, we may turn to more substantive considerations.

PART **II**

The Product:
Some Selected Effects

4

A Strategy for Studying
Second-Order Effects

The Nature of Second-Order Effects

A sixteenth-century explorer who had visited a dozen widely dispersed lands of the Pacific Ocean could not essay a map of that body of water and its lands. He might, however, check out some prevailing myths about the region (e.g., "None of the natives of any of these widely scattered islands possessed those exaggerated ear lobes which are said to account for great acuity of hearing. Indeed they are of average hearing, insofar as we could tell, and the myth seems to originate in the practice of certain of these natives wearing large sea shells as earrings, which at a distance appear to be extensions of the ear.") Or he might pass judgment on the merits of exploration, on whether there was anything that was not already known; or on methods of exploration; and he might generate some hypotheses for future explorers.

Our position is like that of an explorer with a finite amount of time who has selected a dozen such islands for a quick look, a long confab with the chiefs, attendance at some feast, an expert geological survey, and so on. The point is we did not and could not undertake a mapping of *the* second-order consequences of the space program. Our self-assigned task was to conduct what we have called a series of "probes," of pilot explorations in what looked like fruitful and feasible areas. It should be further stressed that we did that for which social scientists are often criticized, i.e, researching the researchable, rather than the "important." We did all this in the expectation that even a modest exercise in data gathering might enlarge or readjust our perspectives. "Important" issues, as perceived by those concerned with them, are frequently not directly researchable, e.g., the impact of space exploration on world peace. Such issues as world peace are multidetermined, and the net impact of a single event is impossible to assess. One can, however, identify specific manifestations of the influence of space exploration, such as the U.S.-Soviet agreement to cooperate on certain matters. In any event, one must begin somewhere.

In this chapter I will concentrate on the perspective with which we emerged from our efforts, using substantive findings for illustrative purposes. Certain broad aspects of these revised perspectives may be mentioned in advance of discussing any of the findings.

The concept of "impact" has connotations that do not bear up under close inspection. It suggests a relatively linear and deterministic causality: in fact, a large social innovation such as space exploration exists in a host system, which spawns and supports it and responds to it as a complex set of demands and opportunities — and responds in a complex and creative fashion, transforming the space program and its meaning in the process. Religious people expand their idea systems to handle the prospect of intelligent life in space, but in so doing they redefine this prospect (though certainly not necessarily the reality) in conventional terms, e.g., pointing, as we shall see, to precedents in the Bible. Communities respond to large aerospace installations not as passive objects struck by an overwhelming force but as highly active systems reacting each in terms of its own needs and resources.

We entered our work with reservations as to the possibility of distinguishing the space program from many other events of our society, particularly from the over-all complex of modern large-scale technology. While in some respects the distinction can be made (e.g., no U.S. programs other than those of NASA and the military are putting men and objects into space), nothing we learned reduced our concern over the possibility or desirability of regarding the space program in all instances as "something different." As a matter of fact, as we proceeded we became increasingly perplexed by the corollary question as to what the space program is.

Objectively considered, the space program is a large set of missions, resources, myths, expectations, activities, and demands. For some people it is power; for some, science; for some, exploration — or prosperity, or economic waste, or madness. We have not been able to chart the full set of these perceptions, but merely to probe some of their complexities. We were forced to conclude that for some people the space program was not so important as a thing in itself but as a sort of indicator of a society's strength in science, education, and so on. It does seem, however, that support for the space program is not support for a single entity, but consists rather of a coalition of different people with different interests each of which finds some manifestation in some one of the faces of a single large social event.

While it is difficult to specify just what the space program means to various people and to differentiate it and its impact from all other events in the world about it, there is little doubt that the space program plays a mythical role that transcends its own actual purview.

We must regard the efforts of NASA itself as representing both a massive innovating technological effort and a symbol of associated technological efforts not directly under its purview.[1]

While many have questioned the extent to which NASA's efforts are at the forefront of science and technology, there can be little doubt but that NASA is the public exemplification of this forefront. NASA has also come to represent the "Space Age revolution," the origin of a sort of rational emancipated man, of discontinuity between generations, of a new age of rationality, with opportunity being rewarded according to ability.

Various mythological roles of NASA will come under our scrutiny in the pages that follow. For example, the myth of a new era of equality according to ability proves vulnerable since all segments of the society do not have equal probability of developing the appropriate abilities. In fact, the "Space Age" may represent a situation in which the Negro's relative ability to participate fully in our society may become poorer rather than better. Here they are mentioned as examples of the extent to which events such as space exploration become dramatized in their public treatment. In considering the railroads as an analogue of space exploration, I mentioned the fact that close historical analysis reveals that dramatization of such important events distorts what goes on. It contributes to the notion of distinctiveness, inevitability, revolutionary change. Our own closer scrutiny of a number of phenomena produced an impression of the new imbedded in the old, of old issues appearing in new context, of new events being absorbed and transformed and playing a relatively smaller role and having a more gradual impact than one might have anticipated from their more dramatized treatment in popular works and the press.

[1] Robert Rapoport, internal memorandum to The Committee on Space of the American Academy of Arts and Sciences (Cambridge, Mass., December 1966), p. 3.

Finally, we will probably leave the reader with a problem of perspective of his own. It may well be that the social scientist, with his emphasis on analyzing things in context and on a comparative basis, tends selectively to stress that which new events have in common with others, and that the social scientist is thereby ill-equipped to identify distinctive, novel developments.

Development of a Strategy

In the conduct of human affairs, our actions inevitably have second-order consequences. These consequences are, in many instances, more important than our original action.[2]

We are concerned in this volume not with the direct consequences of the primary mission of the American space program but with its second-order effects. We may take it for granted that space science will add to our knowledge, that the placing of man in space will add not only to our knowledge but to our ability to explore outer space, and that the building of the hardware and systems associated with the space program will involve spending much money and hiring many men, mostly of rather high degrees of skill and training. Our concern is with the wider ramification of such predictable and understandable effects and, more specifically, with advancing our ability to understand such second-order effects.

What I will comment on in this portion of the book is a series of exploratory probes into the detection of those second-order effects, which reflect a compromise between our conception of what ideally *should* have been done and what could be accomplished with our obviously finite resources of time, money, and personnel.

The early months of our activities revealed the following state of affairs: concern over first-, second-, and possibly third-order effects of space exploration was by no means a new phenomenon. Serious consideration of the effects actually dated back before the launching of the first Soviet Sputnik.[3] This concern consisted mainly of informed speculation on possible effects and pragmatic state-

---

[2] "Space Efforts and Society: A Statement of Mission and Work," a document of The Committee on Space Efforts and Society of the American Academy of Arts and Sciences (Boston, January 1963); see the Appendix, p. 199.

[3] Cf. Donald N. Michael, "Man-into-Space: A Tool and Program for Research in the Social Sciences," *The American Psychologist, 12,* No. 6 (July 1957), 324, for an early statement of a program of research and a plea for base-line data.

ments about research that ought to be done.[4] However, very little data had been gathered. The major exceptions to the paucity of actual data were the existence of a number of polls of public opinion, to be reported on later in this section of the book, and a survey of the attitudes of subscribers to the Harvard Business Review.

The potential agenda of work was enormous, as was observed from the topical outline of the work proposed by the NAS/NRC study group in the summer of 1962, given in full in the Introduction to this book. Not too long after the Work Group on Space got under way, I gave the following, much more discursive version of the outline to an alumni club:

. . . let's sample some of the results which have been seriously proposed as consequences of the exploration of space. They include: changes in man's conception of himself and of God; almost incredible consequences of vastly expanded communications via satellite communications systems (for example, one school of thought suggests that the available t.v. channels would increase so much that a conference such as this would be held with each of us in our offices or homes employing a conference t.v. network); improved short and long range weather forecasting; moment-to-moment surveillance of military installations throughout the world, including virtually immediate detection of hostile missile launchings (I guess maybe this has already happened); contact with beings higher, lower or sideways from us; *or*, if there is no contact, speculation and concern over the possibility of contact; drain on our economy and military strength; *or* stimulus to our economy and military strength; competition with the Russians; *or* cooperation with the Russians; *or* some combination of the two; drain on skilled and scientific manpower; *or* stimulus to the development of skilled and scientific manpower; changes in attitudes toward education and toward stupidity; revolutions in medicine via new knowledge, via telemetry, new substances, and use of computers for diagnostic purposes; revolutions in data processing and retrieval (partially by using communications satellites to facilitate central storage); stimulation of our system of higher education; or disruption of our system of higher education.

I have not even touched on the complexities which some of these developments—many imminent, and some already in being—could pose. Certainly, improved weather forecasting seems at first glance to be an

[4] Cf. Donald N. Michael, *Proposed Studies on the Implications of Peaceful Space Activities for Human Affairs* (Washington, D. C.: U.S. Government Printing Office, 1961); Herbert E. Krugman and Donald N. Michael, eds., *The Journal of Social Issues*, No. 2 (1961), special issue on "Social Psychological Implications of Man in Space"; and "Some Social Implications of the Space Program," in *A Review of Space Research*, NAS/NRC Publication 1079 (1962), Chapter 16, pp. 1–24; and Joseph Goldsen, *International Political Implications of Activities in Outer Space* (Santa Monica, Calif.: RAND Corporation, 1959).

obvious mixed blessing, (but) such forecasts can be made only with improved probabilities, and never with certainty. Which of you would care to assume responsibility of announcing that the prospects for a dry August were .9 in Maine, while they were only .6 in New Jersey? This sort of prediction might well bankrupt the Jersey resorts, and crowd the Maine resorts to the rafters. . . . suppose the people act as though you are right, and the Jersey hotels go bankrupt. Next year, the weather in Jersey is good. What do people do? Sleep in tents on the beach? Or, does the government buy up the bankrupt resorts and reopen them?[5]

In the foregoing passages we may note three levels of approach. The broad topical outline of the NAS/NRC panel gives an idea of the range of issues that might be encompassed. The initial portion of my own passages gives a picture of the sorts of substantive propositions, some of them mutually contradictory, that seemed plausible. My final speculation on the complications that might arise from the presumed benefits of improved weather forecasting affords a taste of the potential ramifications that could be anticipated by a diligent imagination.

In mid-1967 it is obvious that many of the anticipated achievements of the space program have become reality. As I write this manuscript I am smoking a pipe made from a material developed for missile nose cones, sitting next to a television set on which I have watched programs broadcast live from Europe via Telestar. Communications satellites for relaying TV programs now promise such economies that the ownership and operation of them is a matter of controversy. Weather satellites have proved their benefit, giving warnings that have saved life and property. Satellites used for military surveillance are said to have been so successful that they are affecting our strategic relationship to the Soviet Union. Our knowledge of the size and shape of the world, of the temperature of Venus, and of other phenomena in our universe has been altered. And the first men, both American and Soviet, have died in space vehicles.

Yet today, as in 1962–1963 when we were planning our work, there has been little or no research on how people and their institutions have reacted to these developments. What we know is generally what gets reported in the media in the form of newsworthy events but none of the more personal, subtle, and complex responses, which have been the subject of such speculation.

[5] Raymond A. Bauer, "Space Programs: The Joint Responsibility of Business and Government," address to the Harvard Business School Alumni Associations, Baltimore, Maryland, April 9, 1964, Louisville, Kentucky, April 22, 1964.

To be more specific, we reviewed the entries in *The American Behavioral Scientist Guide to Publications in the Social and Behavioral Sciences,* published in 1965, and the *Supplement,* published in 1966. The publications referring to the space program fell into the following categories: "think pieces" of the sort I have already mentioned, including some of the works of our own project;[6] a substantial number of pieces of space law;[7] a limited amount of work on the transfer of NASA technology to the civilian economy;[8] and a limited number of references to public opinion as measured by opinion polls. I shall indicate the highly general nature of these public reactions on the pages that follow and refer to a few additional empirical studies, which escaped the net of the *American Behavioral Scientist.*

In substance, then, I may repeat my statement of a few paragraphs earlier: there was precious little in the way of very solid evidence, nor, with the exception of the field of space law which has built on the established tradition of international law, was there close analysis of available evidence. Just as we decided that the question of the utility of historical analogy ought to be tested by an exercise in historical research, so did we conclude that the utility of empirical research ought to be tested. Our further judgment was that the probability of learning something useful, considering the virgin nature of the issues, would be greater if we conducted a series of pilot probes into a variety of topics rather than concentrating our resources on one or two pieces of work. This decision to spread ourselves thin was re-enforced by our realization of the complexity of the space program as a social innovation and our desire to learn something about its various aspects.

To spread ourselves thin, however, did not mean to be indiscriminate. The term "fishing expedition" is often used with derogation when referring to certain types of research effort. However, the wise fisherman knows where to fish and the types of fish he is likely to catch with the equipment and skills at his disposal.

I may take the category of "economic contribution of space ex-

[6] E.g., John R. Seeley, Bertram M. Gross, Sumner Meyers, Lewis A. Dexter, Edward E. Furash, "Space, Society, and Social Science," a report to The Committee on Space Efforts and Society of the American Academy of Arts and Sciences (Cambridge, Mass., 1963).

[7] E.g., A. F. Haley, *Space Law and Government* (New York: Appleton-Century-Crofts, 1963); and P. C. Jessup and H. J. Taubenfeld, *Controls for Outer Space* (New York: Columbia University Press, 1960).

[8] E.g., J. F. Welees, *The Commercial Application of Missile/Space Technology* (Denver: Denver Research Institute, University of Denver, 1963).

penditures" proposed by the NAS/NRC study group as an example. Three of the topical areas suggested are: multiplier effect of space expenditure on GNP; multiplier effect at regional or local level; and analysis of cost in a one-customer market. These could be regarded as "fished-out" by other people or as amenable to methods at which we were not particularly adept. The fourth of their economic areas, transfer of space technology to the general economy, seemed a fertile one for us to explore.

The problem of technology transfer had been approached prior to the beginning of our work, primarily on the basis of analogy from the past. It was assumed tacitly or explicitly that space-program technology would be useful and that its use would result from the automatic process of civilian businessmen becoming more aware of its utility and adopting it. In the early 1960's there was great awareness of the amount of money going into military and aerospace research and development. Under the assumption that this new technology would somehow find its way into the civilian economy, the primary mode of analysis was a macroeconomic one of analyzing the amount of money going into research and development on the national level. Only the first doubts about the assumption of rapid and relatively full-scale transfer were being raised.

Partially because we sensed the opportunity for a distinctive contribution in this area, and partially because there was available a group of researchers under the direction of Professor Richard Rosenbloom of the Harvard Business School, we probed this problem most deeply of all, concentrating on the process whereby new technology might move from aerospace industry to the civilian economy. This work will be reported on by Professor Rosenbloom in Part III of this book.

Traditionally sociologists have expected that the trends of a national society get reflected on the level of the local community. A number of communities, notably Brevard County, the site of Cape Kennedy, and Huntsville, Alabama, the site of the Marshall Space Flight Center and the Redstone Arsenal, had received massive doses of contact with the aerospace program. Peter Dodd did a pilot study of these two communities to see what a trained sociologist might learn on short order that could affect our views of the impact of aerospace installations on such communities, or on the society of the future.[9]

---

[9] Peter Dodd, "Social Change in Space-Impacted Communities," a document of The Committee on Space of the American Academy of Arts and Sciences (Cambridge, Mass., August 1964), 66 pp.

The interests of a number of scholars associated with the program in manpower problems, and the apparent urgency of the issue, produced a number of reports. On this topic, once more, we felt that other students of the problem had quite thoroughly pre-empted certain topics and approaches. The main topics of concern at that time were in the gross numbers of technically and scientifically trained people required for various programs, including NASA, compared to the numbers of persons presently available and being trained. Our feeling was that somewhere in the social fabric were forces that would modify the conclusions reached on the basis of such gross numbers. We looked at the subculture of research and development in the civilian and the governmental segments of the economy to see whether there were, as alleged, forces which caused migration to government R & D and inhibited the flow from government to the civilian economy, to the detriment of the latter.[10]

Rapoport and Laumann traced the career patterns of technologists who had received their bachelor's degrees ten years previously, and Rapoport, Laumann, and Ferdinand interviewed seniors who were graduating from technological institutions to find out what forces, and what preceptions and aspirations, could be assumed to shape past and future careers of such persons.[11]

During the course of our project, the issue of the role of technicians came to the fore, not only as a potentially key one for manpower supply but one which seemed to reflect a complex relationship to the social structure and possibly offered special advantages for such disadvantaged groups as young Negro males. Two conferences on this subject were called for and reported on, and because of its key nature Chapter 8 of this book will be devoted to the results.

We have referred several times to the two broadly separable aspects of the space program: as a massive socio-technical-economic enterprise and as space science and space exploration. The studies just referred to obviously are pertinent mainly to the first of these

[10] Earl P. Stevenson and John Voss, "The Impact of Massive Federal Programs on the Career Potentials of Scientists and Engineers," a document of the Committee on Space of the American Academy of Arts and Sciences.

[11] Robert N. Rapoport and Edward O. Laumann, "Technologists in Mid-Career," a document of The Committee on Space of the American Academy of Arts and Sciences (Cambridge, Mass., 1964), 60 pp.; Robert N. Rapoport, Edward O. Laumann, and Theodore Ferdinand, "The Power of Choice: Career-line Decisions of Technologists' Class of 1964," a document of The Committee on Space of the American Academy of Arts and Sciences (Cambridge, Mass., 1964), 59 pp.

aspects with its here-and-now consequences. Other studies referred more to the anticipation of the consequences of space science and exploration. Bridging these aspects were a study of businessmen's attitudes toward the space program and a study of the distinctive language habits of aerospace technologists. The study of businessmen's attitudes toward the space program was designed to take advantage of my own earlier study of the attitudes of subscribers to the Harvard Business Review — both to get a relatively rich view of how a key group would respond to the various aspects of the program, and to trace trends in such attitudes during the crucial period of the early years of the program.[12] McNeil's analysis of the language practices of aerospace technologists was an effort to understand the interrelationship of occupation and thinking as reflected in language.[13] Finally, Wiebe made a study of the reactions of religious fundamentalists to the prospects of discovering life in outer space.[14]

We concentrated on the topics where our competency was greatest, the speculation most rampant, the data most absent, and where some data at least, could be gathered. This meant that we specialized in "social" aspects of the situation, in contrast with economic aspects (where considerable work was being done) and political aspects (an area proposed by the NAS/NRC panel but where little data could be gathered).

I will not attempt to summarize the major research writings but merely to draw on them for the broader conclusions which may be reached from these data-gathering expeditions.

The gathering of data cannot be considered as necessarily a good thing. It may be an inefficient use of time, money, energy, and other resources or may produce totally irrelevant information. Grubby empiricism may be a flight from responsible consideration of the implications of already available evidence for the problems with which one has to deal. And so on. Our exercise in data gathering was based on a faith that information is *generally* a good thing, but with little foreknowledge as to what extent and in what ways the information we might gather could be useful.

[12] Edward E. Furash, "Businessmen Review the Space Effort," *Harvard Business Review, 41* (September–October 1963), 14.

[13] David McNeil, "Speaking of Space," *Science, 152,* No. 3724 (May 13, 1966), 875–880.

[14] Gerhard D. Wiebe, *An Exploratory Study on the Implications of Space Flight for Religion,* a document of The Committee on Space Efforts and Society of the American Academy of Arts and Sciences (Cambridge, Mass., 1962), 21 pp.

Furthermore, the purpose of our data gathering was not so clearly defined as it is in much research. Our main interest was not in the testing of individual, clearly defined hypotheses — though some such existed — but rather in exploring certain areas that prior knowledge and social theory suggested would be "interesting." The inelegant way to characterize some of our activity is usually, as I have said, to call it a "fishing expedition." There was method, however, in our choice of fishing locales and methods.

We recognized the dual aspect of the space program both as a massive technical-economic effort and as a program of space science, technology, and exploration. Our own efforts were concentrated mainly on the first of these aspects, because it seemed to us obvious that the major immediate identifiable consequence of the program would be a result of this technical-economic effort. Tangible effects of space science and technology were just beginning to appear, e.g., improvements in weather forecasting and communication. The probability of detecting identifiable reactions to these developments, given our limited resources, seemed remote. However, for the immediate future most of the significance of such extraterrestrial activities will lie in people's anticipation of what will happen.

Most of the issues at which we looked actually constitute observation points from which to consider a series of themes that cut across the substantive issues: the role of the family as a mediating institution and as one affected by sociotechnical innovation; the role of formal institutions such as technological institutions, the church, the community, and so on. The choice of "space impacted communities" as an object of study is an especially obvious example of such an observation point. We reasoned that communities like Cape Kennedy, and Huntsville, Alabama, where the lives of the people are dominated by aerospace activities might in some way represent a microcosm of the future.

In the chapters that follow I will delineate the broad themes that emerged from our adventure in gathering such data.

5

The Public Reaction

I have argued that the space program is a complex phenomenon. It is adventure, exploration, science, technology, rationality, blasphemy, deprivation, displacement of things and people, military power, scientific knowledge, prestige, the future, and so on. The space program wears an infinite number of faces, limited apparently only by the imagination of man to see the potential of so large and complex an enterprise. And these faces stand for the various interests that unite people in support of or opposition to the space program. To understand the faces of the space program and the reaction of people to them would be to understand such broad and general things as the bases of public support, and such a highly specific and directly utilitarian thing as the basis for recruitment of personnel into the program.

A content analysis of news stories and commentaries on the space program would yield a substantial inventory of the things that the space program is seen as being and doing. Yet such an inventory has limited utility since it does not tell us which people hold what view with what frequency and what intensity. It will be my contention in the following pages that press coverage offers inadequate guidance on these questions for reasons that are specifiable and repeatedly observable. I will argue further that systematic opinion polling could offer such guidance, that enough poll data exists to suggest its utility, but that the state of systematic research on this issue is extraordinarily short of what is possible.

Let me first demonstrate the utility of systematic public opinion data. There will be many who feel that, in view of the extensive press coverage of the space program, any diligent reader and viewer of the mass media is adequately informed. The difficulty with regular press coverage is that the press concentrates on issues and controversy. The prevailing practice of news stories is to report the positions of involved persons on both sides of a controversy. This practice has two merits. It generally reflects the feelings and judgments only of people whose feelings and judgments are relevant to the issue and ordinarily does not include "just anybody" who may never have given a serious thought to the issue, and who would be incompetent to do so. Furthermore, the presentation of the range of arguments on an issue is a proper device for educating the persons to whom such messages are addressed. A Congressional Committee, for example, ought to hear all valid arguments regardless of the number of persons who espouse each of them. The job of such a committee and that of the involved citizenry is to reach an informed, and hopefully a correct decision, with its popularity being (again hopefully) a secondary matter.

However, the manner of reporting issues in the news media is often such as to communicate some unwarranted implication of the distribution of sentiments on the issue. For example, by mid-1963, the press carried a considerable number of articles featuring criticisms of the space program. Thus *Business Week*, May 11, 1963, featured a cover story with the title, "Is the Moon Race Hurting Science?" and the *New York Times Magazine* of May 26, 1963, carried a story with the subhead, "Concern is expressed that the space-defense research-and-development program not only is not yielding the 'spin-off' benefits expected but may jeopardize economic growth." Our files accumulated many such stories. Many stories spoke of "growing criticism" or "waning support" for the space program. Of course what may have happened is that the very existence of the space program ceased to be as newsworthy, and the controversial aspects of it more worthy of attention. All that one can be sure of is that a critic or group of critics existed, or a critical argument existed; there is no way of telling what quantity of public sentiment lay behind any of this.

A Case History: 1960 and 1963

Fortunately we were in possession of reasonably definitive data in the form of that rarity, a trend study covering the years 1960–1963 during which it was alleged that support for the space program had lessened. In 1960 I had, with the assistance of Edward Furash, surveyed the subscribers to the Harvard Business Review on their attitudes toward the space program.[1] In 1963 Furash re-surveyed that population.[2] He was careful to make his questions and coverage comparable to those of the earlier study.

It is true that subscribers to the *Harvard Business Review* are not "the public," but neither are scientists, salesmen, newsmen, or any other special group. The point is that it is the *only* group on which comparable and reasonably comprehensive data are available over what appears to be a crucial time period.

It is interesting that data from Furash's follow-up study *could* have been reported in journalistic style to give the appearance of rampant controversy. Here are quotations from four respondents:

The space program is important, no doubt, but far too expensive as of now. (Owner of a soft drink company in Oklahoma.)

The space program has become an extremely important part of our economy. We must forge ahead without wrecking our economy. (Senior vice-president of a Los Angeles investment house.)

I think the space program is the greatest boondoggle since Egypt built the pyramids. (Vice-president and director of a large consumer goods manufacturing company.)

The space program is a well-balanced and timely effort. It is more important to our future than many government ventures costing as much or more but which bring in more votes. (President of a precision instrument manufacturing company.)

Neither the 1960 study nor the 1963 study, properly considered, reflected any of the divergency of opinion suggested by these contrasting, isolated statements.

The major conclusion reached by Furash and myself was that the group surveyed in 1960 was highly enthusiastic about the space program, though not sufficiently informed to permit one to regard that support as stable, whereas the 1963 group was only slightly less enthusiastic but better informed. Hence the 1963 level of support seemed more stable, and one could have more confidence in its enduring.

The most conspicuous possible defection of our businessman sample might have been on the question of the budget for space exploration. In the 1960 study, some 71 per cent of the 2,000 re-

[1] Raymond A. Bauer, "Problems in Review: Executives Probe Space," *Harvard Business Review, 38,* No. 5 (September–October 1960), 6.
[2] Edward E. Furash, "Businessmen Review the Space Effort," *Harvard Business Review, 41* (September–October 1963), 14.

spondents favored increasing the NASA budget, while in 1963 only 30 per cent of the somewhat larger sample of 3,000 would raise the NASA budget, and 49 per cent wanted to keep it where they estimated it to be. It is not clear, however, whether this is a decrease in support. Which is stronger "support" — an opinion to keep spending at a $5 billion per year rate in 1963 or a vote to increase spending over the $900 million rate in 1960?

Perhaps a better guide would be to look at how the NASA program fared relative to other government programs or public expenditures. In both years, businessmen were asked whether they favored the following other areas of spending over space research: medical research, education, new hospitals, cutting taxes, power plants and dams, foreign economic aid, more leisure, and consumer goods. In both years there were strong preferences for medical research (86 and 88 per cent), education (84 and 85 per cent), and new hospitals (58 and 64 per cent) over civilian space activities. It will be noticed that the proportion of support for nonspace activities was trivially higher in 1963. However, while space research lost out to medicine and education in both years, both times it received a majority vote over all the other items. But support for cutting taxes at the expense of the space program went from 27 to 45 per cent, while foreign economic aid dropped from 41 to 29 per cent. In 1963 a tax cut was "in the air," and foreign aid was under fire. It would appear that the shifts between 1960 and 1963 were more in the evaluation of the various yardsticks against which the space program was being measured than in the evaluation of the space program itself.

This judgment is confirmed by the consistency of response on the two surveys to several series of attitude questions, of which I offer a few for illustration. For example, more than 75 per cent of the executives responding to both the 1960 and the 1963 surveys agreed with the following global statements reflecting broad enthusiasm for the space program: (1) "I'd hate to put any limit on what will result from the space programs. After all, anything could happen. Look what has happened in the past" (more than 80 per cent, both surveys); (2) "Outer space is the new frontier. Research and exploration will have profound and revolutionary effects on our economic growth" (more than 75 per cent, both surveys); (3) "Mankind wants to go into outer space because it is there. . . . We are drawn by our desire to know and conquer anew" (more than 87 per cent, both surveys); and (4) "The horizons that will be opened to man by the exploration of outer space are not recognized by most people today" (more than 88 per cent, both surveys). In 1963 the euphoria was slightly tempered but very, very little.

Apropos practical payoffs from space research, from 79 to 88 per cent of the men in the 1963 survey expected new medical and biological knowledge, robot devices, new mathematics and physics, compact nuclear power, and new fabricating materials. What is most significant, however, is that in every one of these instances the level of expectation was higher than it had been in 1960. It is true that in the public arena some sharp questions had been raised about the nature of some claims for economic and technological benefits from the space program; but, for the one group for which we have comparable evidence over the span 1960–1963, there was an increase in confidence in this order of benefit. True, this increase may have been due to excessive claims between 1960 and 1963 that the critics had just begun to attack in 1963, but there are no data to support this theory.

A number of explanations might be invoked to explain the differences in impact created by the mass media in 1963 and that conveyed by our two surveys. I have indicated my own preferred explanation — that the media's accounts of a controversy outline the issues but do not reflect the extent of support for one or the other side.

Having presented what I hope is a reasonably persuasive case for the utility of direct measures of the opinions of various publics, I would like to turn to the question of what we can learn on the various faces of space and the bases for support and criticism of the program.

Interrelationship of the Faces of Space

Some six months prior to the launching of the Soviet Sputnik about one American in five could give interviewers from the Survey Research Center of the University of Michigan a reasonably accurate statement as to what a space satellite was, and almost half (46 per cent) had a vague notion of them.[3] Six months after the Sputnik launching the portion of the American public aware of space satellites had risen to over 90 per cent. Sidney Hollander interviewed Baltimoreans, also six months before Sputnik. Possibly

<hr/>

[3] Data from surveys conducted by the Survey Research Center, University of Michigan, reported in J. M. McLeod and J. W. Swinehart, *Satellites, Science and the Public* (Ann Arbor, Mich.: Survey Research Center, University of Michigan, 1959.)

because he graded his respondents more strictly than had the University of Michigan surveyers, he found only 18 per cent showed some understanding of the term "satellite." However, one week after Sputnik, when Baltimoreans were asked "What do you think is the most important thing that has happened in the last three weeks?" 59 per cent said the most important thing was the launching of Sputnik.[4] Clearly Sputnik was able to capture people's attention.

This was true not only of Americans but of people around the world. A survey conducted in eleven countries of the world by International Research Associates, Inc., found that in Western Europe 90 per cent or more knew about the launching of the satellite, and for practical purposes everyone who knew about it also knew that it had been launched by the Soviets. In English-speaking Canada, Japan, Britain, and Mexico City, the proportions aware of the satellite ran from 70 to 83 per cent. The lowest point was Rio de Janiero, where only 57 per cent knew about the satellite.[5]

Gabriel A. Almond has summarized both American and foreign reactions to the Soviet Sputnik. He groups American reactions into the following areas: (1) foreign and defense problems, (2) American military vulnerability, (3) scientific and technological prowess, and (4) confidence in American leadership and American society.[6]

Speaking of relevant poll data, he says, "These indices of the popular mood in the immediate post-Sputnik period showed a sense of urgency in American opinion. . . ."[7] Americans indicated support for tightening of educational standards with a greater emphasis on science, mathematics, and foreign languages, and more money spent on education. They also showed an increased willingness to support a build-up of military strength.

It would seem most accurate to say that the predominant tendency was for Americans to regard the Soviet Sputnik and America's subsequent failures and successes as an *index* of the over-all state of affairs rather than to regard them as events of importance

[4] The Hollander data is reported in Donald N. Michael, "The Beginning of the Space Age and American Public Opinion," *The Public Opinion Quarterly,* 24, No. 4 (Winter 1960), 575.

[5] The INRA data is reported in Gabriel A. Almond, "Public Opinion and Space Technology," *The Public Opinion Quarterly,* 24, No. 4 (Winter 1960), 556. The poor score in Rio de Janeiro seems to be due to the prevailing proclivity of INRA surveys conducted in this area for USIA to over-sample low income districts in this city.

[6] *Ibid.,* p. 567.

[7] *Ibid.,* p. 568.

in their own right. In the months following Sputnik only 4 per cent of American adults bothered to view it and its successor, and precious little in the way of scientific and factual information about the universe was retained from the flood of information in the media.

For Europeans, it was also presumed that space activities would be taken as an index of scientific, military, and perhaps cultural progress and strength. Almond did a diligent job of reviewing various series of public opinion data to attempt to trace the impact of Soviet and American space programs in Great Britain, West Germany, France, and Italy.

Space successes seemed, between 1957 and 1960, to be a quite powerful index by which to judge the scientific progress of the Soviet Union and the United States. As might be expected, we have no pre-Sputnik data. Immediately after Sputnik, the citizens of Britain, France, and Italy thought the Soviets were well ahead in the race of science, while the West Germans gave us some slight margain of doubt. After our successful launching in 1958, Britain, Germany, and Italy favored us, and France reduced its margin in favor of the Soviet Union. But, by 1960 a survey of ten countries by the Gallup organization found the Soviet Union back out in front in all countries except Greece and the United States. The shifts which Almond reports tend to be substantial. In Britain, where the shift was largest, 58 per cent favored the Soviet Union in 1957 compared to 20 per cent favoring the United States. By the fall of 1958 the proportions were the Soviet Union 30 per cent, and the United States 43 per cent, a net gain of 51 per cent for the United States.[8]

The correlation of the strong shifts in European opinion as to who is ahead in science with events in the Soviet and American space programs indicates that to a large extent space is seen as science, or vice versa.

Space is not so powerful an index of military strength in the public's mind. Various Western European polls of opinion about over-all American and Soviet military strength show no systematic shift between the period immediately after Sputnik and the period after our own success. During this period, however, estimates of Soviet strength in atomic weaponry increased, and sentiment for neutralism increased in Britain, West Germany, and Italy between May 1957 and October 1958.

8 *Ibid.*, p. 557.

Unfortunately, estimates of atomic weapon strength, and preference for a neutral position in the East-West conflict, can be affected by so many events of the world (more than are estimates of scientific advance) that it is risky to make an inference of causation between these trends and space activity. Almond offers us no direct evidence of this linkage, e.g., there is no indication that people were asked such questions as "Why do you feel this way?"

Data on the public's linkage of the space program to military power are very murky. For example, in January 1958, the Opinion Research Corporation asked people, "Looking to the future, what would you say is the real meaning of Sputnik for us here in America?" Michael summarizes the answers as follows:

Twenty-three percent thought that we must catch up in *education, science*, and *defense;* 15 percent saw it as a further threat to our *security;* another 12 percent saw it as evidence that Russia was ahead in *scientific research;* while 11 percent professed to see nothing significant in the accomplishment. Eighteen percent saw it in terms of space *exploration* and the *advancement of science,* 6 percent thought it was a good thing because it shook our complacency, but 5 percent saw it as *propaganda* and bluff. And 1 percent felt that this was an *invasion of God's territory,* and thereby wrong.[9] [Author's italics.]

Clearly so many of these factors are interrelated that, even if the specifically "military" answers had been more clearly isolated, one would still have to contend with the possibility that much of the concern with education and science might have some military component.

Our own research indicates that the interrelationship of the various faces of space is a matter of considerable importance. In both 1960 and 1963 we asked our sample of businessmen to rank five reasons for support of the civilian space program: pure research and gaining of knowledge; control of outer space for military and political reasons; tangible economic payoff; challenge and adventure; and winning the prestige race with the Soviet Union. In both years, science and knowledge were rated in first place, and military and political considerations were rated second, comfortably ahead of the other reasons. One might think the citing of these two reasons an either-or phenomenon, with persons who favored the space program for the reasons that it advanced science and knowledge rejecting the notion that the program should be pushed for military reasons. The opposite is true. The combination of both as

9 Michael, "Space Age and Public Opinion," p. 576.

first and second choices was very popular, with 54 per cent of all 1963 respondents rating the pair in the one-two position.

Furash suggests that analysis of all the data available from our surveys indicates that the "science" and "military" reasons given for supporting the space program are part of an over-all desire to keep our country strong:

> . . . *while most of our executives see the goal of space research as scientific and economic payoff, this goal is also very much a part of keeping the United States strong and in front of the U.S.S.R.*[10]

In the same surveys there is a series of other data that aids our interpretation. In 1960, 19 per cent agreed with the statement, "This whole idea of competing with Russia in a race for space is nonsense," while by 1963 the proportion rose to 28 per cent. Similarly, in 1960, 67 per cent agreed that "The country that controls outer space controls the destiny of the earth," but only 58 per cent in 1963. And, in 1963 we found 71 per cent saying "No" to the question, "Do you think that control of outer space is the most important military objective that our country should have?" (This question was not asked in 1960.) Other answers showed that over the period of three years, American businessmen were drawing a sharper differentiation between the peaceful and the military and political uses of space. They made more informed specific distinctions but did not dissociate the two.

One can therefore conclude that while military considerations were important in their support of the space program, people were realistic about the relative role of the space effort in a total military program and were becoming increasingly so.

A similar analysis of the opinions of the general public is simply impossible because of the paucity of data. We can assume that if the basis for support of businessmen is as involved as our data indicate, that of the general public must be equally complex and less well thought through.

Range of Public Opinion in the United States

Apart from the various possible rationales for supporting the space program, one might want to consider some over-all measure of public support. At no point have any poll data indicated strong general public support for the space program comparable to that evidenced by our businessman sample. (As a matter of fact, an

[10] Furash, "Businessmen Review Space Effort," p. 22.

Opinion Research Corporation study of 1962 indicated that businessmen were the most enthusiastic among all the groups in that study.)

Twice in 1966, in June and then again in September, the Opinion Research Corporation asked the American public, "If some of the federal government programs had to be cut, which of the programs listed on this postcard would you cut first?" In both instances the space program was first to be listed, being selected by 48 per cent in September, in contrast to the second-place choice of the poverty program, 24 per cent. Aid to cities was mentioned by 21 per cent, while aid to education and aid to veterans — down toward the bottom of the list — were cited by as few as 5 and 2 per cent of the people.[11]

In 1962 and 1963, ORC asked people which of many issues required either "urgent" action or "some action but not too urgent." Items such as "reducing juvenile delinquency" were mentioned by 70 per cent in 1963, and "keeping the American economy ahead of Russia" by 62 per cent. "Developing rockets that land an American Astronaut on the moon" was in 25th place on a list of 26. It was mentioned as needing urgent attention by 18%, beating out only "financial support for artists and art activities."[12]

Trend data over time is sparse indeed. In the 1962 and 1963 ORC surveys just mentioned, concern for urgent action on the moon program dropped from 22 to 18 per cent, which might be characterized as "from very low to slightly lower." A longer series is that of the Iowa Poll, reported in the *Des Moines Register*.[13] From 1961 to 1963, the Iowa Poll asked a sample of Iowans whether or not they thought it was important "to send a man to the moon and back safely?" In 1961 the proportion who said it was "very" or "fairly important" was 60 per cent; in 1962, 62 per cent; in June 1963, 47 per cent; and in October 1963, 39 per cent. The high of 62 per cent occurred in 1962 shortly after the United States had completed its first manned orbit. This was also the period in which Iowans were most sanguine about our chances to beat the Soviets to the moon.

These data could reflect a waning interest in the space program. They could also reflect a maturing interest in which euphoria was replaced by more sober enthusiasm. (This is the way in which we interpreted our businessmen's study.) One could also argue plausi-

[11] Data supplied by ORC.
[12] Data supplied by ORC.
[13] "Moon-Flight Interest is Declining," *Des Moines Sunday Register*, January 24, 1964.

bly that opinions on a topic such as this are highly labile, likely to respond to the international climate, the state of the economy, a major success or failure of our or the Soviet program, and so on. One argument that cannot be made, however, is that these data generally represent a high level of support, whether trending up or down, whether momentarily inflated or deflated.

This rather consistently low level of support may be due to the tendency of pollers to present the space program regularly in terms of the moonshot, the utility of which is not clear to all and which is only a portion of the program. Another vulnerability of the space program is that the person interviewed is often asked to rank the space program against other programs whose results are more tangible. However, a Louis Harris poll of late 1965 asked people if they thought it was worth $4 billion a year for ten years to "finally put a man on the moon and to explore outer space and the other planets." Of his sample, 45 per cent said it was worth it, 42 per cent said it was not worth it, and 13 per cent were not sure. Thus, when people were presented with a fuller version of the space program than simply the moonshot it polled a slight plurality. Harris comments: "The reasoning of those who support the program wholeheartedly lies largely in the desire to learn more about the universe."[14]

However, when the same people were asked to rate the space program against nine other programs ranging from national defense to *desalinization of water,* the space program lost against each!

Furthermore, the same people were asked if they would support space spending at its present level if the Russians were not in the race against us. Only 38 per cent would support the present rate of space spending if the Russians dropped out of the race.

A Harris poll in mid-1967[15] showed a drop in support. Whereas 45 per cent thought the space program was worth the cost in 1965, only 34 per cent thought so in 1967.

The most favorable aspect of support for the space program is that it comes selectively from the young, well educated, and wealthy. In the 1967 Harris poll cited above, only 22 per cent of those earning under $5,000 thought the space program was worth what it cost. This proportion rose to 56 per cent among those respondents with an income over $10,000. The 1965 poll showed comparable income differences and the following education differences:

[14] "Public Has Doubts about Space Program," *The Washington Post,* Monday, November 1965.
[15] "Space Programs Losing Support," *The Washington Post,* July 31, 1967.

for those who had grade school education or less, 24 per cent agreed, as compared to 59 per cent among the college educated. Various polls show the young more enthusiastic than the old. This was true among our businessmen. And, the ORC 1963 poll of the general population found that, as one proceeded up the age ladder from the 21–29 year old group to the 60 year and older group, there was a steady decline, from 24 to 12 per cent, of the proportion who thought rockets for the moonshot needed urgent attention.

It may be that we have missed some sizable and valuable poll of public opinion data. However, a canvas of the major polling agencies in the spring of 1967 uncovered nothing additional of promise. If I may assume that what I have presented here is an approximation of our knowledge about Americans' views of the space program over the past decade, then we have a rather typical picture of the limitations of what is characteristically done in polling on public policy questions. First, there is the obvious absence of continuing series in which the same questions are asked over a reasonably long period of time. Stephen Greyser and I have recorded the difficulties of plotting trends in another area, attitudes toward advertising.[16] The same picture repeats itself here and in other areas.

Public Opinion Summarized

Granting the poor data base of our knowledge of public reaction to the space program, we may still summarize the apparent nature of that reaction. There is little doubt that the public has a complex view of the program, though this complexity has not been much explored. It seems equally likely that the public regards the space program as a sort of summary social indicator of our progress in science, technology, economy, education, and so on. I make this statement by implication from the initial public reaction to the launching of the first Soviet Sputnik, which became the occasion for intensive self-examination on the part of the American people, followed by the phenomenon, which Almond reported, of increased public support for defense, science, mathematics, foreign languages, and expenditures on education. This assessment of progress in space exploration as an index of societal accomplishment is offset by the perception of the space program as a discretionary expendi-

[16] Stephen A. Greyser and Raymond A. Bauer, "American Advertising: 30 years of Public Opinion," *The Public Opinion Quarterly, 30,* No. 1 (Spring 1966), 69.

ture meeting no one pressing immediate need. This I infer from the repeated expression of willingness to cut back on space expenditures as compared to other domestic expenditures.

The space program ends up as a low salience issue on which, if we had more adequate data we would almost certainly find that public support is highly volatile, responding to the latest success or failure of us or the Soviets. Evidence of this from another country is found in Exhibit I, a chart of British opinion of U.S.-Soviet standing in the space race over the years 1960–1964. Authoritative sources say that the same trends are reflected in unreleased surveys done since 1964. We find in this chart a gradual secular trend toward improvement in the perception of the U.S. position. What is more striking, however, is the great sensitivity of British public opinion to recent developments in either the Soviet or U.S. space effort.

While our knowledge of the reactions of the general public is poor, that of specific significant publics is even poorer. The reaction of the business community as surveyed by the *Harvard Business Review*, showed that the business community's enthusiasm was accompanied by the expectation of a wide range of economic, technological, and social payoffs. It was my intuitive judgment in 1960 that, despite the fact that they nowhere said so directly, businessmen favored some form of covert economic pump priming.[17] As events have developed, particularly as the economy has edged toward an inflationary phase as a result of the Vietnam War, these attitudes may indeed have changed.

The amount and nature of support for a program such as space exploration by the community of scientists and technologists is acknowledged to be an important consideration for policy guidance. During the 1960's important spokesmen for science and technology have spoken out, many of them critical of the expenditure of funds on the moonshot as opposed to space science, or on space science as opposed to other scientific and technological ventures. It may have been sufficient for policy purposes that articulate spokesmen air their views and that policy makers take them into consideration regardless of the numerical backing of each point of view in the scientific and technical community.

One relatively uninformative poll of the members of the American Association for Advancement of Science was conducted in 1964.

---

[17] Cf. Raymond A. Bauer, "Keynes via the Backdoor," *Journal of Social Issues, 17,* No. 2 (1961), 50.

The direct data gathered by this poll showed that a majority of the members who returned their questionnaires preferred a longer time schedule for the moonshot and a smaller proportion of our national budget on the space program in general.[18] The editor of *Science* commented in his editorial, "At present scientists go along with the space program but without enthusiasm."

Certainly the scientific and technical community is of sufficient importance that we should have a better understanding of their attitudes toward the many faces of the space program. There has been a plethora of statements — before Congress and in scientific and popular journals — by spokesmen for science and technology. But, we know very little about what the rank and file of scientists and technologists think.

Finally, Rapoport, Laumann, and Ferdinand made a small study of the factor that might incline graduating seniors of City U. (a large urban technical institute) and Cosmos U. (a large national technical institute) to see or not to see a career with the space program.[19] Obviously one of the important aspects of such a large enterprise is its ability to elicit support in the form of a supply of competent personnel. About half of the sample of seniors from both institutions felt that a NASA career was acceptable or strongly favored it. Enthusiasm was greater, however, at the lesser institution, with about 19 per cent of the City U. graduates saying that a NASA career was strongly favored, as compared to about 8 per cent of the seniors at Cosmos U.[20]

The attraction of a NASA career lay in the "science" image of the Space Agency and the space program, but enthusiasm was less among the cream of the crop of graduates, specifically the graduates of Cosmos U., many of whom looked forward to graduate work and an academic career. The unfavorable elements were "cut-backs, excessive bureaucratic control, over-specialization and consequent danger of obsolescence, immersion in something too large in which one becomes a nameless, faceless functionary, and so on."[21]

Thus, the attractiveness of a NASA career is balanced off between the positive aspects of a "science" image and certain conditions of employment associated with large government enterprises,

[18] *Science, 145* (August 7, 1964), 539.
[19] Robert Rapoport, Edward Laumann, and Theodore Ferdinand, "The Power of Choice: Careerline Decisions of Technologists' Class of 1964," a document of The Committee on Space of the American Academy of Arts and Sciences (Cambridge, Mass., 1964), 59 pp.
[20] *Ibid.* Data interpolated from Table I, p. 11.
[21] *Ibid.*, p. 10.

the weighting of this balance being modified by the alternatives one sees before him. The very best students, having more alternatives in prospect, are a little less favorable.

It appeared that students perceived other dimensions of a NASA career on the basis of their personal psychological orientation. Using an experimental procedure for psychological screening, Rapoport and his associates classified their students tentatively into four categories:

Rationalist: main rewards come from mental effort.

Conventionalist: looking for safe, tried and true lines of development.

Pragmatist: orientation is in terms of making things work.

Activist: oriented to doing things that are socially significant.

While the technique was experimental and the classification tentative, the procedure did seem to identify persons with meaningfully different career aspirations, e.g., persons heading for a career in science were more than twice as likely to be classified as "rationalists" as were persons who intended to become engineers. One third of the latter were classified as "pragmatists," a category into which *none* of the budding scientists fell. What is of interest to us, however, is the kinds of reasons each of these types gave for finding a NASA career attractive. The quotations offered are illustrative rather than representative, and are intended only to suggest the fruitfulness of this line of approach:

Rationalist: "I'd like to work for NASA on research. They support research for the sake of research. It's hard in business to find this kind of thing."

Pragmatist: "We have to get ships into space. It is impossible for private companies to do this if we are to enforce our anti-trust laws, so it must be the government."

Activist: "I feel the desire to do something important . . . I feel that I would owe a certain amount of my time and effort to others, . . . with something that was doing something important to the nation—like NASA."[22]

The "Conventionalists" were least interested in a NASA career, and the authors offer no example of their attitude toward NASA, except to say that they "tended to think favorably of the idea in terms of job opportunities, fringe benefits, and so on. . . ."

I conclude that our exploratory work with samples of businessmen and graduating seniors from technical institutes indicates the

[22] *Ibid.,* p. 30.

possibility of a meaningful approach to the multidimensionality, the many faces, of the space program.

Yet, the multidimensional approach is contrary to the established tradition of journalistic opinion polling, which has dominated our thinking on the sources of support for public programs. For reasons of economy, effort, ease of asking questions and of communication to the public, opinion pollers have regularly sought a single "thermometer" type of question, which can be used as a summary statement of the *amount* and *trend* of support. In the case of the space program this has usually taken the form of asking for approval or disapproval of the moonshot, an estimate of whether we or the Soviets are "ahead," or whether expenditures on space exploration are or are not favored over expenditures on other programs. With enough such data and enough daring, meaningful inferences of a limited sort can be made from the answers to such questions. However, it is difficult, even with trend data from questions such as this, to explore the multiple bases of support and opposition to something so diverse as space exploration.

It is clear that national and public agencies have not been particularly interested in the sorts of questions we have raised here. If they had been, there would be more data. It is therefore fitting to close this section by raising two logical possibilities. Perhaps the topic is not of sufficient national interest to warrant more data gathering. Or, maybe there is not sufficient faith in the utility of such polling information to generate support for gathering it, even though the topic may be of great importance.

6

Mediating Institutions

Public opinion, or the opinions of various publics, may be thought of, in one sense, as "effects," or, in another sense — as we did in the previous chapter — as the bases for support of the space program. Probably all "effects" occupy such dual status, since any program will be supported in terms of the effects it is seen as producing. However, other more concrete consequences, such as those we will consider from this point on, are closer to what is generally meant when we talk about second-order consequences. In this chapter we deal with aerospace towns, and the family and other institutions that mediate the impact of the space program on the larger society.

The Aerospace Town

Perhaps the most dramatic, or at least the most visible and tangible secondary effect of the space program has been the creation of what we might call the aerospace town. There is nothing new in a town being rapidly expanded or even being established *de novo* as a result of some larger activity — a mining town, a railroad town, a town with a newly established military post, or more recently towns established or selected for Atomic Energy Commission activities.

Such new or rapidly expanded communities have a number of things in common, among them an increase in money to be made and people to be serviced. Generally, there is not only an increase in the total amount of money available, but usually the newcomers are better paid than the local inhabitants and frequently, but not always, better educated. The services required fall partly in the private sector of the economy, e.g., food, lodging, clothing, and personal services, where they represent an opportunity for the established community to make money, and partly in the public sector, notably education, where they are often regarded as an imposition on the established community. Some services, such as health, social work, and recreation, may be undertaken by private, voluntary, or public agencies. The influx of newcomers may be seen as a stable, long-term proposition or as a temporary matter to be exploited while the opportunity is ripe. There may or may not be a power struggle between the old-timers and the in-migrants.

My task here is not to present an exhaustive taxonomy of the response patterns that may occur when a program such as that of NASA establishes an activity in a relatively small town. It is rather to exploit the fact that, there being a few towns on which the space program has had considerable impact, the experience of these towns may teach us something about the space program and its personnel. Especially one may raise the question as to whether in some sense these towns offer a prototype of "the future," it being assumed that the local community is an observation point of manageable size from which to view the patterns of forces working themselves out on the national level. If that assumption is correct, then it is reasonable to look at a few towns dominated by the aerospace industry to see what prospect of the future they present.

To test the plausibility of this proposition, Peter Dodd visited a number of cities in the aerospace crescent reaching from Cape Kennedy to Houston and spent relatively more time, in the summers of 1963 and 1964, in the two key areas of Brevard County, Florida, the residence community for the Cape Kennedy installation, and the noted space center at Huntsville, Alabama, where activities spearheaded by German scientists under Werner Von Braun have now grown vastly beyond their original size. Here he interviewed community leaders and others, consulted records and written materials, and did the various things an ethnographer does.

The gross story of these two cities can be told in a few statistics. Huntsville grew from 16,000 in 1950 to 80,000 in 1960 and passed 100,000 by 1964.[1] Brevard County grew even more rapidly, from 23,000 in 1950 to 220,000 in 1965.[2] The educational level of both

[1] Peter Dodd, "Social Change in Space-Impacted Communities," a document of The Committee on Space of the American Academy of Arts and Sciences (Cambridge, Mass., August 1964), p. 7. The 1950 and 1960 figures are from the Census; the 1964 figure is a local estimate.
[2] Charles Grigg and Wallace A. Dynes, "Selected Factors in the Deceleration of Social Change in a Rapidly Growing Area" (Tallahassee, Fla.: Institute for Social Research, Florida State University, September 1, 1966), p. 7.

communities increased markedly. "The median number of school years completed was 12.1 in both Brevard County and Huntsville in 1960; in Florida as a whole it was 10.9 and in the United States 10.6. The change for Brevard County, 1950–1960, is from 10.2 to 12.1, a remarkably high increase."[3] These figures can be taken as indicators of the magnitude of increase in the size of the population, of the demand for services, and the change in composition of the population, which became not only a more highly educated one, but a wealthier and younger one, with a higher proportion of children.

Add to the picture the fact that both communities are located in the South, the reputed heartland of social conservatism, and one would anticipate a good deal of social dislocation and difficulty topped by a power struggle between the established elite and the new, more highly educated, wealthier newcomers — probably focused on demands for better educational services.

It would be incredible if two communities expanding as rapidly as these did not have some difficulties. There were such difficulties; and each of the communities handled them, sometimes in contrasting styles. However, the difficulties were so much less than anticipated, and the absence of a power struggle so marked, that Florida State researchers were prompted to discover a systematic explanation which they set forth in the "jargonesquely" titled paper, "Selected Factors in the Deceleration of Social Change in a Rapidly Growing Area."[4]

The ability of a community to meet demands of rapid expansion is, of course, a function not only of the magnitude of the demands, but also of the community's resources and its disposition to use those resources to meet the demands. The probability of a power struggle, in turn, is a function not only of the unmet needs of the new constituency, but also of the disposition of that new constituency to seek power.

Huntsville and Brevard County varied in their capacity to serve the needs of the newcomers. Huntsville was a single well-integrated community, whereas Brevard County was a collection of smaller communities subject in total to a larger impact. As a result, two

---

[3] Dodd, "Space-Impacted Communities," p. 12.

[4] Grigg and Dynes, "Deceleration of Social Change." For more detailed studies of the Cape Kennedy area see other reports of the Institute for Social Research, Florida State University: "A Report on the Institutional Capacity of Three Communities in the Cape Kennedy Area — Titusville, Cocoa, and Melbourne" (January 1965); "Community Satisfaction and Community Involvement in Brevard County" (January 1965).

contrasting patterns developed. Huntsville was able to rapidly expand and maintain its own services — police, welfare, education, and so on — to meet the new needs. Its established voluntary institutions, such as the churches, were able to absorb the needs for social service posed by large numbers of mobile families.

In 1964, the City of Huntsville Planning Commission reported that "there are relatively few slums or sub-standard houses in its corporate limits. . . . The majority of houses are less than 12 years old." Informed persons reported that there was very little truancy or delinquency in the city. And, there was concern for neither a social welfare program, nor an active social work agency.

Brevard County found that its established institutions were unable to cope with the demands on them. A Brevard County official in close touch with families in difficulty said of the churches: "Churches are our weakest institutions; when families are in trouble, the last persons we hear from are the ministers." There was also a great deal of complaint about substandard housing. As for social services in general, a Citizens Survey Committee reported in 1963: "The outstanding characteristic of the social welfare program in Brevard County is that services have failed to keep pace with the fantastic growth in the population." A local official reported a high divorce rate in the county.

The failure of established institutions to handle many social problems on an informal basis pushed Brevard County to the development of community resources to handle them in an institutionalized, formal, and legal fashion.

At first glance, the Huntsville experience might seem to be the preferable one, because that city was able to handle the transition with less dislocation. However, if one holds a widely shared picture of what the future will be like, the Brevard County experience may provide for a more rapid transition into that next stage. Certainly the demand (and support) for formal social services to take over the handling of problems formerly handled by the extended family and currently handled by the smaller nuclear family, neighbors, and voluntary organizations represents a rationalization of certain social functions.

Furthermore, in situations such as these there are always opportunists with a cause to serve who wish to take the chance to strengthen that cause. In Huntsville it would appear that established institutions were able to strengthen their position. In Brevard County the transition offered opportunity to the proponents of change.

This can be seen most clearly in the school systems of the two areas. Both school systems were expanded without too much community dissension, since, as drastic as the expansion in enrollment was, the income of the communities increased even more rapidly. However, in Huntsville the school system expanded within its established framework and in a traditional fashion. The Huntsville school system of the mid-1960's reflected little of the new look in curriculum development, not even the new experimental mathematics and physics.

Brevard County was another matter. Its high school was referred to in a national publication as ". . . probably the most unusual, complex, and exciting public school in America . . . Melbourne High."[5]

For the Superintendent of Schools in Brevard County, the influx of aerospace personnel represented not so much a set of demands as an opportunity.

The space program, as such, opens up new opportunities for the schools. . . . It enables us to do more experimentation. . . . The parents are researchers and willing to experiment in the schools.[6]

Here we may notice two factors at work. One is the relative incapacity of Brevard County to encapsulate expansion within the established institutional framework. The other is the symbolic status of the space program as sanctioning broad change and experimentation. In all our deliberations there is no clearer instance that the space program, *in its symbolic role,* produced an impact definably different from any of the other phenomena with which the space program might on one basis or another be classified. The symbolic role of the space program is so clearly appreciable because the image held by the school superintendent was so much at variance with the facts in Brevard County. Aerospace employees in Brevard County were in no sense "researchers" as a group, and they were not even particularly high-level technologists. Yet the image of the space program is such that the school superintendent was able to exploit the myth of NASA as "research" to bring about the experimentation that he wanted. One can assume that the aerospace workers of Brevard County, having been paid the compliment, were loath to reject it. They, too, became subscribers to the myth.

What is at stake is not simply that there were differences be-

[5] *Newsweek, 60,* No. 15 (October 8, 1962), p. 108.
[6] Dodd, "Space-Impacted Communities," p. 36.

tween Huntsville and Brevard County due to the relative capacity of the community to absorb change within its institutional framework, but that given the balance of demands and resources, a preferred pattern seems to develop. One pattern favors smoothness of transition for the immediate future. The other may facilitate transition into the more long-range future. I will return to this question of "the future" later.

Of more immediate concern is the issue of the relative lack of conflict within the two communities. In Huntsville, one might attribute it to the greater adequacy with which the problems of the newcomers were handled. If that were an answer for Huntsville, it could not be for Brevard County, where there were certainly enough issues to warrant a community power struggle if the newcomers had been disposed toward one.

Most observers, including ourselves, seemed to take two things for granted: first, that the newcomers, being more highly educated, and being outsiders, would have values considerably different from those of the local population, and second, that the newcomers, being more highly educated, cosmopolitan, and so on, would be activists who would try to implement their own values. Our expectations proved largely wrong on both counts.

Popular writing, combined with Northerners' stereotype of the South, created the impression that the Southern space communities were being inundated with Northerners. (Where else would one find so many trained people?) A popular writer wrote of Huntsville, ". . . 75% of the population represents immigrants from all parts of the country, particularly the northern states. . . ."[7] NASA's personnel chief in Huntsville told Dodd he estimated that 50 per cent of their new employees came from residences outside Alabama, which of course means 50 per cent came from within the state. However, when Dodd examined the records of the birthplaces of Huntsville NASA employees, he concluded "not more than 20% were born outside the South, broadly defined."[8] Dodd gathered the impression that the population of Brevard County was also largely Southern in its origin. This was subsequently confirmed by the Florida State researchers whose survey showed that well over half of the migrants into Brevard County had in fact spent most of their lives in the South.[9]

[7] Eric Bergaust, *Rocket City, U.S.A.* (New York: The Macmillan Company, 1963), p. 59.
[8] Dodd, "Space-Impacted Communities," p. 13.
[9] Grigg and Dynes, "Deceleration of Social Change," Table 6.

These data indicate that the assumption of value conflicts between newcomers and old-timers might have to be modified substantially in light of the fact that most of the newcomers were themselves Southerners. Furthermore, most observers were probably like ourselves in assuming that conflicts of values might center about race relations. If one were to look at the local community as a microcosm of the national community, and see the space program as symbolizing equality of opportunity, rationality in employment, and so on, one could envisage a struggle to carve out a role for the Negro in "the future" of America.

This did not happen. First, Dodd, as well as Grigg and his co-workers, found no conspicuous militancy on the part of the newcomers. Second, certain issues were being approached outside the community context. The Civil Rights Act of 1964 opened up public accommodations, and the need for Federal funds opened up the schools. There was no occasion for fighting these issues on the local level. Last, and most poignant, was the decline in status of the Negroes in the aerospace communities studied by Dodd.

Despite the Equal Employment Opportunities order, space centers in the South have hired very few Negroes, primarily because qualified applicants cannot be found. As farm land is taken up for the space center and the support contractors, Negroes may actually be displaced from their jobs as farm laborers and move elsewhere in search of work. The size of the Negro community remains the same while that of the white community multiplies. As a result the percentages of Negroes in the total community declines substantially. In Brevard County, it has decreased from 25% in 1950 to 11% in 1960; the figures from Huntsville are comparable.[10]

Add to this the fact that though there was a general increase in income and education in the aerospace communities that of the Negroes remained about constant.

The median family income for Huntsville in 1960 was $6300. For Negro families in a typical Negro district, the median income in 1959 was $2272 and by 1964 was estimated at $2700. For all of Huntsville, the median number of school years completed was 12.1; in the Negro district it was 7.3.[11]

In the light of these data, and other findings of our research, one must question another assumption about the myth of "the future," namely, that the space age will open up opportunities for all.

[10] Dodd, "Space-Impacted Communities," pp. 20–21.
[11] Ibid., pp. 22–23.

Hence, some sources of potential conflict did not exist. Furthermore, our second main assumption that more highly educated, more cosmopolitan newcomers would opt for power in the community proved as erroneous as any of the assumptions we have already examined.

In both Huntsville and Brevard County the newcomers proved markedly uninterested in politics. A Florida county official commented:

The new people who move in are not especially interested in the community and are unwilling to take political responsibility and to participate in voluntary organizations. There are no patterns for them to be active in. If they try to act, they meet a lack of response to their efforts to set up such patterns . . . . For instance the community could sponsor local forums on community problems and in this way keep politicians under public scrutiny, but there is none of this.[12]

Dodd comments on Huntsville that "the newcomers have also been slow to take active part in politics," even though "the churches and the United Fund rely heavily on the participation of technologists from the space center and the support contractors."[13] In other words, when they become involved it is to *serve* the community rather than capture control over it. The same pattern is manifested in the fact that they willingly vote in favor of school bonds but show little interest in running the P.T.A. or otherwise influencing the educational system.

There are obvious superficial reasons for the failure of aerospace technicians to become involved in local affairs. Immigrants into the community require a certain amount of time in which to become known and effective. The involvement of men in their work may deflect their time and energies away from community concerns — though this need not be true of their wives, who also do not seem to get involved. However, there is evidence that lack of political involvement runs deeper than these superficial reasons.

In Brevard County, the migrants were found to have been much less likely to vote even in national elections. Only 23 per cent of the old-timers failed to vote in the 1960 presidential elections, compared with 43 per cent of the early migrants and 52 per cent of the recent migrants.[14] In other words, nonvoting in the presidential election was twice as frequent among the migrants, and a migrant

---

[12] *Ibid.,* p. 27.
[13] *Ibid.*
[14] Grigg and Dynes, "Deceleration of Social Change," Table 3, p. 144.

was about as likely to stay home as to go to the polls. Some of the nonvoting among the recent migrants may have been due to ineligibility because of inadequate time of residence, but this cannot account for the nonvoting of the early migrants. When one considers the educational level of the migrants, this is an extremely strong apolitical orientation.

Holmes reports a similar pattern in Los Alamos.[15] Los Alamos is a town more totally dominated by technologists than either Huntsville or Brevard County. It was built almost entirely from scratch two decades ago with no old-time resident population to run it. For a long period of time the voting turnout was well below that for the state as a whole. Only recently have the residents of Los Alamos begun to vote at a rate consistent with their educational level. Even at that they show a disposition not to bother about things that are running well. For example, they vote in favor of school bonds but do not bother to turn out in numbers for school board elections. Only 20 to 25 per cent vote in these elections. Holmes comments:

> The stable rates of turnout for school elections may be but the measure of voter's satisfaction with their schools . . . (I have little doubt that any fundamental change or threat to those schools would produce an awesome uproar . . .).[16]

We have here in part an explanation of the lack of community conflict in the aerospace communities — the newcomers show little disposition to run things or express a voice in public policy — and in part a comment on "the future" to the extent that aerospace families represent an important segment of that future.

The aerospace technologist and his wife are apolitical both because of the husband's strong involvement in his job, and because of the family's mobility. It would appear that positive commitments are to the job, and to a nonlocal community, which Dodd refers to as "alumni of place and time," friends and acquaintances that they have known in other times and places.

These "alumni of place and time" may be viewed on one hand as a source of alleviation of the strains placed on the families which have had to move frequently, and as an additional unanticipated source of reduction of community conflict. They not only perform

---

[15] Jack E. Holmes, "Science Town in the Politics of New Mexico" (Albuquerque, New Mex.: The University of New Mexico, Department of Political Science, 1967).

[16] Ibid., p. 19.

some of the social-work functions that community institutions might otherwise be called on for, but they serve as a reference group, a focus of attention. The prototype of "alumni of place and time" might be found in modern America among the peacetime military. And, in fact, Dodd has the following to say:

> The best example of "alumni of place and time" was given by the son of a retired West Pointer who is now a technologist employed at Cape Kennedy. Questioned about his relatives, he reported that he seldom saw his uncles and aunts and knew them only slightly. Later, in talking he mentioned his "Uncles" and "Aunts" often enough that I asked about them. The uncles were his father's West Point classmates.[17]

I will discuss the role of these "alumni of place and time" when I consider the role of the family among aerospace technologists. Here, I introduce them — found among the nonmilitary as well as the military aerospace families — as a further factor turning aerospace families away from local and national political concerns.

One must assume that if residential patterns become more stable, these families will take more roots in the community, and will take part in running it. The pattern to date, however, does not re-enforce the myth of an elite group that is in all respects eager to run the society according to a new set of values. Furthermore, the experience of Huntsville and Brevard County indicates that to the extent that other groups do a satisfactory job of running things the "new elite" will be willing to let them do so.

The Family and Other Mediating Institutions

Although families are to be found in all societies, there are great variations both in their internal structure and their relation to other social organizations. In agricultural and pastoral societies the basic family unit is the "extended" family. In the most primitive societies the family may be the only significant organization, with larger groupings composed of clusters of extended families. With the growth of social development, other organizations come into being, particularly governments, armies, churches, and trading groups. . . . With the development of industrial civilization, these organizations tend to detach themselves from family influence. Concurrently, the extended family begins to break down and the basic family unit tends to dwindle into the "nuclear" family.[18]

In several of our studies the American family showed up as both

---

[17] Dodd, "Space-Impacted Communities," p. 53.

[18] Bertram Gross, "Social Systems Accounting," in Raymond A. Bauer, *Social Indicators* (Cambridge, Mass.: The M.I.T. Press, 1966), p. 195.

the cause and the effect of occupational choice, success, and activities.[19] The formation of the nuclear family to which Gross refers is regularly regarded as the natural response of the family to the demands of an industrial society. The economic functions of the extended family are to a large extent taken over by business organizations. The need to move about in order to take advantage of opportunities dictates that the size of the family unit be kept to a minimum — husband, wife, and children. If a large unit is involved either more interests have to be sacrificed to a given move, or the disposition to move has to be stifled. And, finally, many of the services provided by the extended family are now taken over by specialists — laundry and cleaning, emotional support and settlement of conflicts, care of the sick and aged, aid in financial crises, and so on.

But there are considerable functions remaining within the nuclear family. In addition to the procreation and physical rearing of children, the family transmits the values and norms of the major culture, and the variants in those values and norms of the subculture to which it belongs. It provides role models for boys and girls. It provides not only the physical resources but the motivation and the cognitive map of the world that affect the educational and occupational chances of its younger generation. And, the ways in which it goes about these tasks can vary considerably, as can the outputs it produces in the way children grow up to take their place in the adult world.

The chain of relationships affecting the final result is complicated by the magnitude of the phenomenon and can be illustrated by Table 1, taken from the work of Rapoport and Laumann in studying the occupational history of men who had been out of school for ten years.

[19] These studies include Dodd, "Space-Impacted Communities" and other documents of the Committee on Space of the American Academy of Arts and Sciences:

Robert Rapoport, Edward Laumann, and Theodore Ferdinand, "The Power of Choice: Careerline Decisions of Technologists' Class of 1964" (Cambridge, Mass., 1964), 59 pp., a study of the career planning of seniors of two large technological institutes, "City U." and "Cosmos U."

Robert Rapoport and Edward Laumann, "Technologists in Mid-Career" (Cambridge, Mass.), a study of the background and career histories of ten-years-out graduates of three technological institutes, "City U.," "Upstate U.," and "Cosmos U."

Laure Sharp and Raymond A. Bauer, "The Future of the Scientific and Engineering Technician," a report of two conferences held in Washington, D. C., Summer 1964 (Cambridge, Mass., 1964).

TABLE 1

Type of Position Held and Father's Occupation*

(per cent distribution)

| | Father's Occupation | | |
| | Professional | White Collar | |
Type of Position Held	and Business	and Manual	Total
Scientific and highly technical	37	30	34
Managerial and others	34	23	31
Lower level technical	29	41	35
Total	100	99	100
	(N = 211)	(N = 184)	(N = 395)

* From Robert Rapoport and Edward Laumann, "Technologists in Mid-Career," a document of The Committee on Space of the American Academy of Arts and Sciences (Cambridge, Mass.), Table 5, p. 19.

While these data are not especially strong, they certainly indicate that even after one has achieved a bachelor's degree his life chances are affected by the sort of family he grew up in. We know, in addition, that there are many data showing the extent to which the very opportunity to get higher education is related to family background. There is no news in announcing that American children of different social strata have different life chances.

However, in general thinking, this has usually been attributed to such factors as the cost of getting educated. The data from our studies suggest more emphasis on factors related to personal motivation, and particularly to the relative emphasis placed on commitment to profession and family by family of origin. Social historians have labeled the work-oriented ethos in Western culture, the Protestant Ethic. While there has been a great deal of ethnic mixing in American society, and adoption of a general American culture, even so crude a measure as that of family religious background shows, throughout the work of Rapoport and his colleagues, that the Protestant Ethic has some remaining validity in distinguishing one group from another.

Table 2 shows the religious background of students at three technological institutions, City U., Upstate U., and Cosmos U. As the names are intended to suggest, these three institutions represent three levels of increasing prestige, of quality of faculty and students. One set of figures dominates the table. Cosmos U. gets 28 per cent of its students from homes with a Catholic background. City U.

gets 42 per cent. Since they are located in the same city, the differences cannot be explained by differences in geographical access.

TABLE 2

STUDENTS BY RELIGIOUS ORIGIN IN THREE UNIVERSITIES*

(per cent distribution)

Religion Brought Up In	City U.	Upstate U.	Cosmos U.
Roman Catholic and Eastern Orthodox	42	33	23
Jewish	12	10	15
Protestant	44	50	55
None, other, no answer	03	07	02
	(N = 124)	(N = 159)	(N = 145)

* From Robert Rapoport and Edward Laumann, "Technologists in Mid-Career," a document of The Committee on Space of the American Academy of Arts and Sciences (Cambridge, Mass.), p. 13.

Two of the correlates of a successful career orientation are a long time perspective and a realistically high career expectation. Rapoport, Laumann and Ferdinand asked the graduating seniors of City U. and Cosmos U. when they thought they would be at the peak of their careers and what their anticipated income would be at this peak. Of the seniors from City U., 63 per cent thought they would reach their peak in *less* than twenty years, and 61 per cent thought they would be earning *less* than $17,500. Among the seniors from Cosmos U., 61 per cent thought they would reach their peak in *more* than twenty years, and 71 per cent thought they would be earning *more* than $17,500.[20] These differences hover around the proportions of 2 to 1. The experienced data analyst may ask why we do not make the comparison on time perspective and anticipated income directly on the basis of religious background. It is because these data were collected at the end of four years in a specific educational institution, a point at which exposure to a given institution as well as family background (which affected the probability of going to that institution) have had an effect. The data for controlling simultaneously on religion and institution are not available to me. What I shall concentrate on is the portrayal of two populations descriptively different on certain important dimensions.

Another correlate of a high professional orientation is willingness to move, or perhaps this is a measure of the relative emphasis placed on family as compared to work. And perhaps, mobility is also a function of opportunity. One might argue that highly com-

20 Rapoport, Laumann, and Ferdinand, "Power of Choice," pp. 18–19.

petent technologists receive more offers requiring that they move out of their home locale. But one might also argue that having more offers would permit them to live where they choose. However, whatever refinements and complications one may want to introduce, it would appear from the entire range of evidence available to us that willingness to move (which might be phrased inversely as a lack of social, economic, and psychological ties to one's own locale) is a key correlate of success in the new, highly technological society, and especially in the volatile aerospace industry — just as it was during the period of any frontier of history.

Let us take a closer look at the graduates of City U. and of Cosmos U. (Upstate falls in the middle in all instances, and the use of only the two extremes makes comparison easier.) The students at City U. come from families of lower socioeconomic status, whose perspectives are more limited (e.g., they are less likely to distinguish the relative merits of City vs. Cosmos, and have a less elaborate notion of career options). They are of ethnic and religious backgrounds more associated with an emphasis on family than on career, and in fact they are more likely to marry before finishing school (43 per cent at City compared to 24 per cent at Cosmos U.). Furthermore, the curriculum is based on the cooperative plan, a circumstance that may establish a tie to a local firm before graduation. Cosmos, on the other hand, is — in addition to being at the opposite end of the continuum on the preceding variables — a national institution, attracting students from all over the world. And, since a much higher proportion of the graduates from Cosmos go on to graduate school, they may be forced to make a move even before entering the job market.

There is, then, a strong alignment of factors suggesting that the graduates of Cosmos U. ought to be more willing to move than the graduates of City U. Operating against this is the fact that the city in which both are located is a very attractive job environment for technologists. Cosmos graduates are held in very high esteem and are likely to be able to remain in the area if they choose. That is to say, few will move because of very dire necessity.

We find that among the ten-year-out graduates, 80 per cent of the graduates of City U. were born in the state, compared to 28 per cent of the Cosmos graduates. Of these graduates, 52 per cent of City U. graduates were still working within the state, compared to 32 per cent of the Cosmos graduates.[21] Rapoport and Laumann developed

---

[21] Rapoport and Laumann, "Technologists in Mid-Career," p. 29.

an index of mobility, which included the mobility of both spouses with respect to place of birth, education, and current job. The data for the extremes of the scale are presented in Table 3. I include Upstate U. in this instance to substantiate my earlier assertion that it regularly falls between the other two.

TABLE 3

GEOGRAPHICAL MOBILITY IN RELATION TO UNIVERSITY ATTENDED[*]
(per cent distribution)

| Mobility Pattern | University Attended | | |
	City	Upstate	Cosmos
No mobility	48	35	23
.			
.			
.			
Maximum mobility	16	21	39
	(N = 112)	(N = 133)	(N = 117)

[*] From Robert Rapoport and Edward Laumann, "Technologists in Mid-Career," a document of The Committee on Space of the American Academy of Arts and Sciences (Cambridge, Mass.), p. 37.

For the methodologically minded, the probability of these differences occurring by chance are less than one in a thousand. For the practically minded, it is worth noting that the differences between City and Cosmos are easily of the ratio of 2 to 1.

Regardless of the explanations I have invoked, descriptively we have two populations of technologists, one of which is half as mobile as the other. In an era in which work in the aerospace industry places a high emphasis on mobility, one group seems to be more "with it" than the other. And, in fact, we do have data indicating that the graduates of Cosmos are presently more likely to work in large firms involved with a good deal of government work. This does not mean that the City U. people are irrelevant to the aerospace effort. They are precisely the type of technologist needed in the middle and lower levels of large technological enterprises such as NASA; and the City U. seniors are about twice as likely to say that they regard with "strong favor" the prospect of a career in NASA (19 per cent compared with 8 per cent).[22]

If we assume that most of my interpretations of the reasons behind this difference are correct, we may draw two inferences:

[22] Rapoport, Laumann, and Ferdinand, "Power of Choice," interpolated from data on p. 11, Table 1.

(1) family considerations, and background factors filtered through the family, reduce the availability of a portion of the technologist population to Space Age activities; (2) the Space Age is putting an additional strain on an already nucleated family structure. I will take up the first point immediately and the second one later.

It is impossible to say that the antimobility disposition of a portion of the technologist population constitutes an important practical problem. There are no clear data I know of. We do have evidence from our study of technicians — technical people of a subprofessional rank — that nonmobility is a problem with that group. Several authorities told us that technicians are sufficiently nonmobile that trained technicians are essentially a local rather than a national resource.

In the case of technicians, then, it looks as though background factors leading to decreased mobility do restrict our ability to use a meaningful national resource. Hence, we may have an economic problem worth considering with respect to graduate technologists. We may also have a social problem. Though space installations, once established, remain fairly well fixed for the workers it tends to be a project-oriented life. That is, individuals become tied to projects rather than to the installation and move on when the project is finished though the installation remains stationary.

If a disposition toward mobility becomes an increasingly important criterion for participation in advanced technological enterprises the already considerable advantages of graduates of elite institutions may be compounded. The better technologists (though not necessarily the best, many of whom will have academic appointments) may become a stratum of rootless mercenaries. This may not only have consequences for the social structure of the society of technologists, but also isolate this stratum of technologists from participation in the affairs of the broader society, i.e., the picture we found on the community level might become accentuated. This is certainly a question on which data could be gathered and trends plotted over time.

But, what of the effects of the Space Age on the family, rather than vice versa? I have argued that demands of family and other social ties and career demands come in conflict, and those who are more free to liberate themselves from local ties and are willing, relatively, to subordinate family needs are more likely to move in response to career demands. It is, therefore, no more than turning the coin to its other side to say that work in the new aerospace industries and communities places strains on the family.

Dodd,[23] and Grigg and Dynes[24] tell essentially the same story. Frequent moving produces certain rather obvious consequences for the family. Any remaining vestiges of the extended family — ties with parents and other relatives — are attenuated. Children must leave friends and shift schools. Wives must leave friends. While husbands leave friends too, the work place offers them a ready reservoir of potential friends. The family must learn the mores and institutions of a new community. Even shopping requires some learning.

Even though existing social institutions, such as the church and school, social agencies designed to serve specific social-work functions, and alumni of time and place, take up some of the slack, the fact is that the family is faced with burdens of adjustment that it would not have if it remained put, and it is deprived of certain gratifications of living in a familiar place. Of course, the family also has a wider variety of experiences, the wife and children are "broadened," and all that.

Such consequences of frequent moving have been noted in the case not only of aerospace personnel, but of the military, and of junior executives in nationwide corporations. And, the existence of an "alumni of time and place" has been noted in all instances.

The reader should feel free to speculate on the good or ill of such consequences. There is little doubt that frequent moving produces specifiable difficulties. There is probably just as much reason to believe that some unspecified portion of children, wives, or families are in some way strengthened by the experiences they have, though there is no means presently known to me how to weigh such plausible benefits against such probable difficulties. My guess is that most people will emphasize the difficulties, but this may be because they are easier to specify and identify.

There is one "Space Age" development that surprised the researchers. Dodd suggests that the functional demands of the aerospace technologist's activities produce an isolation of work from family. To some extent, particularly at Cape Kennedy, emergency periods often make the man unavailable as husband or father. Furthermore, Dodd reports that there is little discussion of the husband's actual work, in part for security reasons, and in part because the technologists find the job of explaining what they do excessively difficult. In addition, there are heavy travel demands on many of the men.

[23] Dodd, "Space-Impacted Communities."
[24] Grigg and Dynes, "Deceleration of Social Change."

The wife of the engineer-administrator mentioned above said that her children were greatly upset by their father's absence. To this remark another engineer's wife rejoined that she was upset, because her children were no longer aware of their father's absence.[25]

Not all these factors operate in all instances, and Dodd is suitably cautious about the hypotheses he puts forward. He suggests two propositions that are worth investigating. One is that the unavailability of the husband for many of his customary duties is placing an additional amount of responsibility for running the family on the wife, creating perhaps a mininuclear family in which the father is a part-time but necessary member. As a corollary of the first of these propositions, the second is that the father is serving or will serve as a less effective role model than he has in the past. This may cause problems of identification for the boys in the family, a problem of modern life true also for nonspace families in which the father is absent a good deal.

But into all this speculation one may throw one conspicuous bit of fact: for all of the difficulties that aerospace families *may* be having, they seem to be conspicuously healthy. The following is the summary statement of a study made of the health of the families of Space Center employees at Cape Kennedy.

The selected measurements as presented for Space Center employees, their wives, and residents of Brevard County show lower indices of psychophysiologic symptoms, lower death rates, lower measures of illness absenteeism, taller stature, and similar or slightly lower prevalences of overweight and elevated blood pressure than US populations of comparable age-race-sex composition.

These results were not anticipated by us prior to the study. We anticipated potentially deleterious health consequences to rapid population growth and mobility; to absent extended family supports; and to multiple, repeated deadline-meeting stresses, characteristic of the space age's activities. It would appear that these activities and changes were either not perceived by employees as stressful, or the stresses were not manifested in dysfunction measured by our indices, or they were compensated by other, concurrent, situation advantages.[26]

We may then summarize our comments on the role of the family as follows:

The nuclear family does indeed seem appropriate to the Space Age, and to almost any form of technology-based future society we

[25] Dodd, "Space-Impacted Communities," p. 53.
[26] H. A. Tyroler, "Community Health Impact on the Space Industry," *Arch. Environ. Health,* 14 (February 1967), 256.

can reasonably imagine. To the extent that one's family of origin holds only vestiges of the old extended family, one is likely to be hindered in his pursuit of a career by emphasis on early earnings, and by a bias against geographical mobility. I would not care to argue the merits of a family-oriented vs. a career-oriented value system, but merely to point out the demonstrable consequences of each. Not only will the career-oriented youth be more successful, but the work demands of Space Age jobs will probably bring about an increased nuclearization of the family. Perhaps there will be some higher level of synthesis, such as the alumni of time and space.

Is there reason to worry about the consequences of the life of the aerospace family for the family's health and happiness? Dr. Tyroler's study indicates that the inhabitants of Brevard County are an incredibly healthy lot. In addition to their physical health, they score very well on such a traditional indice of morale as absenteeism. Since any source of stress on a complex population is bound to produce some identifiable difficulties, and may also produce some plausible but usually hard to substantiate benefits, one must end up asking how difficulties and benefits balance out. If Tyroler's broad measures of physical and psychological health may be taken as such a summary, one would have to conclude that either the results are not too detrimental, or some Darwinian process had weeded out all the weaklings before Tyroler made his study.

7

Space and
Man's Imagination

One hears and reads that the exploration of space will "broaden our minds," "stretch our imagination," "give us new vistas," and all that. Like many other, but not all, statements about the secondary effects of space exploration, this one, as stated, can scarcely be wrong. I say "scarcely" because it is conceivable that some new phenomenon with which man is confronted may prove so threatening and/or overwhelming that he will panic in the face of it, retreat to more primitive intellectual processes, and show a constriction of the imagination. The space program does not appear to pose this degree of threat for any perceptibly large portion of the population, hence it can be presumed generally to stretch, stimulate, and probably — though not necessarily — improve our imaginations. However, our research uncovered systematic tendencies, out of inertia or *some* degree of anxiety, to constrain its impact on imagination.

If mankind grapples with a new problem, something is bound to happen to man's mind. But mankind does not have a single common mind except in that it has a shared culture. The interesting questions ask, "*whose* minds, and shaped in *what ways?*" When we turn to data to address this question we have, as usual, a number of small pilot studies, no one of which establishes any point very firmly, but all of which complement each other. There were three studies done for our project: children's space comics; "space-speak," the language of aerospace personnel; and reactions of religious fundamentalists to the prospect of life in space.[1]

---

[1] Dorothy Lipson, Robert Rapoport, and Francis Dahlberg, "Children of the Space Age," a document of The Committee on Space Efforts and Society of the American Academy of Arts and Sciences (Cambridge, Mass., September 1964); David McNeil, "Speaking of Space," *Science, 152*, No. 3724 (May 13, 1966), 875–880; Gerhart D. Weibe, "An Exploratory Study on the Implications of Space Flight for Religion," a document of The Committee on Space Efforts

Two other studies were published by scholars, not associated with us, one of children's drawings of satellites, and the other on space jokes as indicators of attitudes toward space.[2]

It may be said in advance that there are some prevailing themes throughout these works, though each does not appear in all of them. The most pervasive is that novelty, to the extent that there is any, is embedded in the usual. Familiar themes appear in space dress, space themes are interpreted in familiar terms, and so on. On the whole, the net impact of this work is that one gets exposed to more that is familiar than to that which is novel. The second major theme, and one which may explain the first, is that man uses his imagination to exploit space for solving certain orders of problems that are pretty much imbedded in his ordinary way of life. Space may (or may not) pose new problems, but it also offers an arena in which to mull over old ones. The challenge to imagination is there, but often the greatest achievement of imagination is to find an old way to solve new problems, or a new pretext for airing old ones.

The Language of Space

McNeil, in his analysis of the language of space technologists, begins with the general observation that in any highly innovative field there is great pressure for a new vocabulary with which to refer to concepts or entities that have not previously existed, to handle new refinements of old concepts or technical versions of nontechnical concepts, to make communication more efficient, or, perhaps, even to set oneself apart from others.

The opportunities for imaginative creation of new words are many. Popular are the metaphor, the metonym, and the neologism. Metaphors, which depend upon observing some similarity between the new entity and some older, more familiar one, produce such space words as "pad," "abort," and "umbilical." Metonyms depend upon contiguity between the phenomenon to which a term refers explicitly and the phenomenon to which the meaning is to be transferred. Thus, "eyeballs in" and "eyeballs out" refer to extreme acceleration and deceleration and are intended to stand for their

---

and Society of the American Academy of Arts and Sciences (Cambridge, Mass., September 1962).

[2] Rhoda Metraux, "Children's Drawings: Satellites and Space," *Journal of Social Issues,* 17, No. 2 (1961), 36–42; Charles Winick, "Space Jokes as Indicators of Attitudes Toward Space," *Journal of Social Issues,* 17, No. 2 (1961), pp. 43–50.

SPACE AND MAN'S IMAGINATION

conditions of movement as reflected in the astronauts' physical state. Neologisms are new words made from elements of old ones, e.g., "rockoon," a rocket launched from a balloon.

In some quarters it is an established practice to poke fun at all jargon. However, the minimum that must be granted coinages such as those just referred to is that they demand considerable imagination to create, and until the figures of speech become so absorbed into the language that the original imagery is dead, they provide an enriched pattern of associations for their user and for the listener and reader.

At the hard-core of space-speak are the nominal compound and the acronym. The nominal compound is basically a phrase in which the order of words has been reversed and the prepositions deleted. Thus, the "vehicle for launching" becomes "launch vehicle," and grammatically has the status of a single noun. Two words tied together are the minimum for a nominal compound but there is no upper limit. One made up of 13 words has been found in the Congressional Record: "liquid oxygen hydrogen rocket powered single stage to orbit recoverable boost system."

One of the established rules of language is that the most frequently used words are short ones. We see one process whereby this occurs in the acronym, the word formed from the initial letters of a phrase, e.g., Nasa, or Unesco. Here I have rendered the words with only initial caps (rather than NASA or UNESCO) to underscore that they have in fact become single words, proper nouns. Whereas many acronyms came about in attempts to shorten long descriptive phrases, the use of them is so well established that "in some cases the sequence is apparently reversed: the acronym is devised first, then a compound is found to fit it" as in the case of EGADS, or "electronic ground automatic destruct sequencer" (the system used to destroy a malfunctioning missile after it has been launched).

Obviously an imaginative acronym such as EGADS, connoting the dismay that accompanies the destruction of a defective missile, enriches our language. But it is an exception. Few acronyms are so rich in connotation, e.g., Unesco, Nasa, and so on.

The main secondary contribution of nominal compounds (beyond their original function of communication) is to add to the pretentiousness of the language. Since originators of jargon have been accused of doing so for their own status enhancement, and in turn have defended themselves by saying they are providing a means of communication, McNeil sought to investigate this issue directly.

He devised a "pretension index," a measure of space program originated compound overuse. This index was applied to samples of writing by NASA engineers and to writing in a popular magazine on space technology. In the popular magazine there were 220 per cent more five-word compounds and 330 per cent more six-word compounds than in the NASA material. A comparison was also made of spoken material of NASA engineers in Congressional testimony and the speech of Congressmen, with similar results. It would appear that the NASA engineers used nominal compounds more parsimoniously, and others, by implication, used them not only for direct communication but to establish their "in" status. The originators of the jargon are at least partially vindicated. They apparently use it mainly for its primary purpose of direct communication.

It is a commonplace that the space program has added to our vocabulary. What is unresolved is whether space-speak will stimulate us in other areas of our language behavior to imitate the richer imagery suggested by the metaphors and metonyms, or whether the plodding, pedantic, pretentious use of nominal compounds will be stimulated. Of course, there is always the possibility that the metaphors of the space program will eventually die, their original meaning be lost, and we will be assaulted by phrases such as "a real down to earth multi-stage launch pad for getting the program airborne."

Space in Visual Art and Humor

Metraux's study of children's conception of space satellites[3] offers one lead to expanded horizons; more specifically, in attempts to depict a satellite in space, children abandon the very concept of the horizon, which she says:

. . . has such a wealth of meanings for us, as it may stand for the future . . . , or for expanded human capacities . . . , or for future accomplishment. . . . Psychologically, the ability to perceive and produce horizons is related to our conception of maturation.[4]

She finds that when children depict a satellite in space they use a multidimensional space with no concern for the limits of the horizon.

Rather, the satellite, the earth and other planets, the sun, the moon, and

[3] Metraux, "Children's Drawings."
[4] *Ibid.*, p. 42.

the distant stars co-exist in a multi-dimensional relationship to one another and to the viewer.[5]

What shines through her description of these drawings, however, is that while they reflect some basic idea of what actual satellites look like, the drawings reflect well-known age and sex differences in cognitive style. For example, boys concentrate on depicting movement, and girls on ornamentation. And younger children focus on the satellite itself without any surround, while older children produce more complex drawings in which the satellite is imbedded in a surround.

The satellite is, characteristically, globe-shaped, but sometimes it is attached to a rocket or has a following rocket; it may have a flowing fiery "tail." Little "waves" (of sound?) come out of the antennae; various machine-like bits are attached to the surface or are visible in the interior.[6]

While the younger children might attach a "beak," "tail," or "eye" onto the satellite, the satellites of the older children are thoroughly mechanized.

Metraux speculates on the possibility that the concept of the horizon that comes about through socialization in our culture is a limiting one, and that the ability of older children (in the post-horizon-acquiring stage) to be freed of this concept while contemplating a space satellite suggests that we may find ways of making this multidimensional freedom more pervasive.

Her data speak to me more powerfully on a quite contrasting point, namely, the extent to which prevailing cognitive styles, associated with age and sex, exert themselves when a person attempts to grapple with a new phenomenon. This is nothing more than a variant of the process involved in projective techniques for psychological diagnosis. What is new may exert some constraints on what one does with it, but it nevertheless offers a good deal of latitude of interpretation that will become progressively constricted as group and personal norms for dealing with it are established. In other words, that which is novel may provide not only demand and opportunity for innovation, but at some level of generality *must* evoke the imposition of old structures and modes since man cannot deal with anything — new or old — without giving structure to it. And, at some level of abstraction, this structure *must* be a familiar one.

[5] *Ibid.*, p. 41.
[6] *Ibid.*, p. 39.

The interplay of the new and the old is well exemplified in space jokes. For example, a group of Martian space men return home, having captured a pair of gasoline pumps. They proudly announce to their leader that they have returned with a pair of earthlings to be used for breeding purposes. The court biologist snorts: "Don't you see, the fools brought back two males!"

Winick analyzed space jokes collected by him and a colleague, Elliott Horne, during the period 1957–1959 in the New York metropolitan area.[7] These jokes were ones told to them spontaneously in social situations and do not include space jokes told by professional entertainers. Winick's summarization is as follows:

> The effects of these jokes are perhaps best understood in terms of what people do with the jokes rather than in terms of what the jokes do to people. . . . Almost the majority of the jokes collected express people's interest in assimilating space into contemporary frames of reference, with relatively little adjustment and a minimum of adaptation. Although the country's scientific effort is concerned with space travel and we know nothing about space people and practically nothing about space travel. In many languages the word for "stranger" is the same as the word for "enemy," and people want to make the space traveler as little of a stranger as possible. . . . People translate space travel into contexts with which they have already come to terms. They thus magically bring themselves closer to coming to terms with this new development and stave off the kind of readjustment which any innovation brings. . . .[8]

He then adds that some space jokes, a small proportion, are antiminority-group stories, i.e., that space is used as a pretext for working out old prejudices.

Winick takes a functional approach to space humor and sees in it a purposive device for assimilating new and anxiety-provoking phenomena into a familiar context, and a pretext for venting old prejudices in a new concept.

There are some, of course, who might enter reservations about Winick's assumption that space jokes represent a "method of coping with the unknown," and might contend that they are merely intended to be funny. However, considering the complex sociopsychological functions that have been attributed to humor in all societies, we must grant Winick at least the privilege of advancing his psychodynamic hypothesis.

Lipson, Rapoport, and Dahlberg[9] analyzed the content of space

[7] Winick, "Space Jokes."
[8] *Ibid.*, p. 48.
[9] Lipson, Rapoport, and Dahlberg, "Space Age Children."

comic books and studied the readers and nonreaders of space comics in two Boston area schools. Their study of readers and nonreaders of space comics focused on the types of questions that readers and nonreaders had about space. By and large, the questions of the readers were somewhat, but not markedly, more sophisticated in that they asked more contingency questions, in the form of "What would happen if . . . ?" and exhibited more intrinsic interest in space, asking such questions as "Why was space made?" The differences are small, and their nature is such as not to demand much comment.

More surprising, however, are the results of the content analyses of the space comics. In some respects they are very much like Winick's space jokes in that they are populated predominantly by men. There seems to be a consensus that space is a man's world. Space comics, like space jokes, are very much concerned with space beings, with space "aliens" appearing in about 75 per cent of the stories. In one third of the stories they are indistinguishable from humans, in one third they are monstrous.

But the themes are the old familiar themes.

In space-science fiction earthmen and people from outer space tend to take on the stereotyped roles of hero and villain respectively. They then reflect the patterns of comic strip characters generally in their goals and in their means to those goals.[10]

And, as might be expected, the villains, i.e., the "aliens," are more likely to use violence, while the heroes, the earthmen, are more likely to use intelligence and knowledge to win.

Winick's interpretation of space jokes as a device for reducing anxiety over space discoveries seems very applicable to the content of space comics, in which the victory of earthmen over space men is a source of reassurance.

The authors complain of the paucity of imagination in space comics.

Men are clothed, like Flash Gordon, in garments resembling dancers' leotards. Skirts are often added for women . . . usually reminiscent of medieval costume. . . . Space-science comic book artists seem to use history, rather than imagination in clothing their characters.[11]

In addition to relying on tradition as a source of inspiration for garb, space-science comics rely as might be expected on our cul-

[10] *Ibid.*, p. 19.
[11] *Ibid.*, p. 21.

tural heritage for themes. Lipson, Rapoport, and Dahlberg identify several recurring ones. There first is that of reversal of roles, in which the youngest child or the most worthless person turns out to be the rescuer (10 per cent of the stories sampled). The next is the "early tale" theme characterized by the presence of a group of local friends who are taken care of by a protector who works "magic"[12] (28 per cent of the sample). A third is the "invincibility through doubling or multiplying the hero" in which heroes are endowed with dual identities, one of whom has special powers (22 per cent of the sample).

Space and Religion

If it seems plausible that the prospect of life in space can evoke anxiety, it has seemed most plausible that the prospect of intelligent life in space might be a source of great anxiety to the religious fundamentalists, who have a fixed view of God, man, man's creation, man's life after death, and the nature of the universe.

In the fall of 1962 Wiebe[13] discussed the topic of life in space in writing and during a luncheon conversation with eighteen leaders of the Southern Baptist Seminar at Boston University, one member of which arranged for the distribution of a small questionnaire to fifty parishioners of two Southern Baptist churches. Ordinarily one would regard information from such samples as no more than suggestive because of the small size of the samples and their unknown bias. However, the great homogeneity of their responses suggests that these people may indeed be representative of the large religious fundamentalist community.

They are perhaps epitomized in the reply of one man who wrote:

> The questions underscored by our pin pricks in space have not knocked this layman's religious world out of its Christo-centric orbit. Instead, John 10:16 has an exciting new meaning for me, and I can see in it more than ample comfort for the orthodox. The King James version of John 10:16 is: And other sheep I have, which are not of this fold; them also I must bring, and they shall hear my voice; and there shall be one fold, and one shepherd.[14]

Note once more the weaving together of the old and the new, the use of neo space-speak, "Christo-centric orbit," and the con-

12 *Ibid.*, p. 9.
13 Wiebe, "Space Flight and Religion."
14 *Ibid.*, pp. 3–4.

structive reinterpretation of the Bible to extend the concept of "other sheep" — which has generally been assumed to be a reference to Gentiles — to include possible space men.

Of the fifty people answering Wiebe's questionnaire forty thought we probably would find some form of life in space, fourteen thought it would be intelligent life, two thought there might be intelligent beings *without sin,* and two considered it probable that if such intelligent beings without sin were discovered they might be favored over human beings in the sight of God. Out of fifty, forty-seven rejected the idea that space flights would bring the astronauts closer to heaven, and only four thought space flights might arouse God's anger. It is intriguing that three of these four were in the youngest age group, thirteen through sixteen. Only five out of fifty thought space flights should be curtailed from "a religious point of view," while eleven thought they should be curtailed from "an economic point of view."

In sum, the more sharply the possibility of a conflict between the space program or space exploration and religion was posed, the more categorically the issue was rejected. And, the opposition came almost entirely from the very young. All five who thought it was desirable to curtail space flights for religious reasons were under sixteen years of age. The older people, whom one might have thought of as more conservative, apparently had a better-integrated view of their religion and were better able to encompass the possible discoveries of the space program within their framework of belief. While one might be tempted to discount this pattern on account of the small sample size, our interpretation is bolstered by the independent testimony of the religious leaders with whom Wiebe talked prior to gathering the questionnaire data: "Our friends the Baptist officials assured us repeatedly that the old anthropomorphic concept of God is rarely found among their people except the children."[15]

In respect to all of these studies (language, children's drawings, jokes, religion, and comics), I have stressed the interaction of the old with the new and the relative surprise of many of the researchers at the relative predominance of the old. The exercises in fantasy were interpreted in a large part as an effort to absorb the new into the old so as to reduce its impact, rather than as an effort to expand one's horizons. Or the arena of space was depicted as a new place to play old games.

[15] *Ibid.*

All of this raises two questions: What are legitimate expectations as to the amount of novelty to expect? And to what extent does the approach of the researcher bias the perception of novelty?

Apropos the first of these questions, it seems to me that perhaps one finds as much novelty as one can legitimately expect to find. Certainly in each of the areas explored there is evidence that people have had to cope in some way with a new problem. Was it a trivial feat for our religious fundamentalists to interpret Christ's reference to "other sheep" to include possible space men? A people are entitled to employ their fantasy as a way of reducing their anxiety about that which they have already imagined. Furthermore, as we have pointed out, some structure must be given to that which is new, and any structure that is invented must in some broader context rely on older structures. In my discussion of Mazlish's work on historical analogy I referred to his conclusion that complex social inventions such as the space program and the railroad tend to get dramatized. I think this is appropriate to our consideration of the degree of novelty and imagination stretching one can expect to find in the mental products associated with such enterprises. The dramatized version suppressed the bread and butter version. Close and relatively systematic study must perforce produce a perception of more that is usual than does a treatment that is tactically committed to establishing the distinctiveness of the phenomenon.

The same line of reasoning carries over to the answer to the second question. To some extent the mode of analysis of the social scientist is relatively biased toward the usual. The journalist asks "What's new?" The social scientist asks "Within what established framework of analysis can I understand things?"

Hence, it seems to me that not too much emphasis should be placed on the relative balance of old and new in the mental products we have studied, but on the inevitable interaction between the old and the new. No social invention moves onto an empty stage of history. The people and institutions of the times have their problems and their resources. They are bound to reflect their problems in their reaction to new events, and they cannot, as I have said several times, come to grips with these new events except by ordering them at some level to old concepts and dealing with them with methods that are in part based on old ones.

8

Manpower: The Special Case
of Technicians

(Written with Laure Sharp)

We have observed how man has adapted himself to the new frontiers, but his adaptability has had its limits. Training is a prerequisite most people could not foresee, or did not foresee, as a prelude to the Space Age. This has left the nation with an enormous manpower problem. A government official told us: "There is urgent need for careful planning (for technical manpower). . . . The Vocational Education Act, for example, says we have to train in relation to need. And if we are going to do this, we have to do some planning."

But, how to do it? Planning is dependent, as we know, on information and on understanding the nature of problems. This chapter is an exercise in the analysis of one of the subproblems of manpower planning: the role of technical support personnel in the manpower picture. The reasons for focusing on this particular subproblem will evolve as we proceed. However, it can be said that in the mid-1960's it seemed to be a pivotal issue around which might revolve the seemingly urgent problems of technical and scientific manpower and other problems of social policy. The period of time covered in this chapter is from 1963 to mid-1965, the time in which we studied this problem. The method of exposition is that of a narration of the evolution of our perspective as we wrestled with the issue.

The Over-All Picture

In the early years of the 1960's there was a prevailing belief that we were faced with an impending serious shortage of technical and scientific personnel. In mid-1963 the Bureau of Labor Statistics completed a report for the National Science Foundation,[1] which concluded that during the decade 1960–1970 there would be a need for a million new scientists and engineers and that the projected supply was fewer than 765,000. This report forecasted a deficit of 265,000 such persons by 1970. It was expected that the situation would be most severe by the mid-1960's, when the engineering graduates would be at their lowest point in years.

In the midst of this picture was the space program. NASA's own staff of scientists and engineers had risen from 3,000 in 1960 to 9,000 in January 1963, while the number of scientists and engineers working on NASA contracts in private industry rose from 5,000 to 34,000. It was projected that this total of engineers and scientists associated with NASA programs would rise to between 90,000 and 100,000 by 1970.[2] This would mean that in 1970 the space program would requisition the services of about 5 per cent of an anticipated 2,000,000 scientists and engineers.

Remarks such as the following, made to us by distinguished scholars in 1962, were not unusual:

NASA is in vicious competition for people with very particular capabilities. . . . Apart from the considerable amount of publicity and expenditure provided by NASA, there must be other areas of human endeavor of equal or even greater social importance that are running short of manpower.

[NASA] must help determine the extent to which its own program may be depriving wider sections of the economy of suitable technical manpower.

However, it would seem that while the actual projected demand for scientific and technical manpower by the space program — 5 per cent of the total — was large, it was scarcely of such an order as to deserve much credit or blame for the over-all manpower picture. In summarizing an overview of the situation for the American Academy project in the summer of 1963, Laure Sharp made this evaluation:

One's first impression—both in terms of the total numbers involved, the specific skill needs, the basically unfavorable employment situation in the aircraft industry, the geographic impact of NASA activities—leads

---

[1] *Scientists, Engineers and Technicians in the 1960's — Requirements and Duties,* prepared for the National Science Foundation by the U.S. Department of Labor, Bureau of Labor Statistics (NSF 63-34, Washington, D. C.: U.S. Government Printing Office, 1964).

[2] These figures are from the 1964 *Manpower Report of the President* (Washington, D. C.: U.S. Government Printing Office, March 1964), pp. 173 ff.

one to suspect that on balance the program as now projected need not have a destructive impact on the scientific manpower situation. Certainly there would be shortages or imbalances even if the program did not exist. The need for new and imaginative approaches to the manpower question, and a re-orientation of our thinking about education and work would be imperative even if we decided now . . . to give up all thoughts about landing on the moon.[3]

Yet, as in so many other instances, the space program directed widespread attention to this issue. Although the demand for technical manpower by the military R & D program was many times larger, there was relatively little suggestion that it might be cut back or kept down. The attention centered on NASA came as a result of factors we have had cause to comment on at various points. The space program was highly visible and highly dramatized. Furthermore, we are confronted once more with the fact that it was seen as a discretionary activity. Relatively few people (though there were some) were ready to suggest that we had the option of cutting back military R & D. Many more questioned the wisdom of the space program.

NASA was sensitive to its role. For example, in 1962, it began a program of grants for predoctoral training with the aim that as the program expanded it would eventually produce 1,000 Ph.D.'s per year. Furthermore, NASA developed its own programs of training and retraining technical personnel. Special emphasis was placed on using and training technical support personnel — technicians — who could reduce the need for fully qualified scientists and engineers.

By the spring of 1964, a number of individuals and institutions came to regard the potential use of technical support personnel with special interest. Not only did the prospect of increased numbers of technicians promise to make the use of full-fledged scientists and engineers more efficient and thereby reduce the threat of shortage, but the technician's careers were seen as a device for testing another problem with which the nation had become preoccupied: that of underprivileged youth, especially Negroes.

There were mounting bits of evidence that the new technological era might breed its own distinctive crop of orphans, those youths who did not have the opportunity and/or ability to get full-scale training at the college level. Unemployment among young Negro men, for example, had grown very much more rapidly than

[3] Sharp, "Scientific Manpower Problems."

unemployment among white youths. At the same time there were pockets of underprivileged, undereducated white youths in similarly precarious situations. This was the year that the War on Poverty began, and the linkage between that program and the possibility of developing a cadre of middle-range technical specialists was readily perceived.

The appeal of achieving a simultaneous solution to two pressing problems was strong, yet serious interest in the many questions surrounding the use, training, and recruitment of technicians was just beginning. Before continuing with the issue of technicians, however, we shall return to a further consideration of the over-all problem of scientific and technical manpower.

Some Problems of Anticipating the Supply and Needs for Scientific and Technical Manpower

The projected shortage of engineers and scientists served as the first stimulus for looking at the potentially expanded use of technicians. Hence, consideration of these projections is an appropriate preface to the more intensive look at technicians.

The problem of anticipating the supply of and need for scientific and technical manpower is the same as that of anticipating any future state of the society, with the added complication that a much longer lead time is involved than is required for most issues on which we might want to take action. At a minimum, if a Ph.D. in physics is required in 1970 a student must choose his area of concentration during his sophomore year of college, and he ought to have made that choice in 1964. But, this vastly understates the case. As we look backward in time, as we shall when we get into the discussion of technicians, we find that events occurring before the child enters the elementary school have a strong effect on whether or not that child some fifteen or sixteen years later has the ability or inclination to become a physicist. And, as we look forward in time, we find that the content of scientific and engineering jobs is changing so rapidly that it now appears it is not sufficient to think of training a man only for his entering job. Specialists concerned with the problem are coming to talk more and more about the need to look beyond the entering job to the need for training scientists and engineers more broadly in fundamental disciplines of science and technology, so that they may change as the job does. Some of this broadening of training is in fact already taking place.

As the government official quoted at the beginning of this chapter noted, the requirements for personnel and the requirements for training are interdependent. Some of the burden is thrown back on the training function because precise requirements cannot be predicted very accurately far into the future, since the content of scientific and technical jobs can be counted on to change in an unpredictable fashion. To the extent that training is directed at developing flexibility in the trainees, *precise* prediction becomes less essential. Anticipation of the range of circumstances with which a future scientist or engineer will have to cope is sufficient. But this can be carried only to a point, since neither the training system nor the trainee can carry the burden of preparing for an infinite range of future requirements.

It is ironic to contemplate the difficulties of coordinating supply and need for highly skilled manpower in the light of fictional versions of genetic selection and early conditioning put forth by the authors of such utopias and counterutopias as *Brave New World* and *1984*. Such proposals imply a reasonably accurate notion of manpower requirements a generation in advance. A careful reading of *Brave New World* shows that Aldous Huxley deliberately posited a *stabilized* society so as to insure a fit between supply and demand. While such speculations may be dismissed as idle fictionalizing, they have in fact been the stimulus for serious discussions of how one might plan and execute a manpower program at some time in the future. It would appear that such practices would be possible only if social policy were adapted to suit the practices rather than vice versa. They do, in any event, underscore the fact that public policies can either complicate or simplify the problem of projecting supply and demand.

The frustration of attempting to make such projections in the face of shifting policies is reflected in the statements of the men confronted with the problem. Colm and Lecht comment:

To a very large degree, the growth in employment of scientists and engineers has been the result of public policy decisions reflected in government expenditures.

These decisions sometimes introduce discontinuous changes in the demand for different programs and objectives. The space program in the early 1960's supplies an illustration.[4]

---

[4] Gerhard Colm and Leonard A. Lecht, "Requirements for Scientific and Engineering Manpower in the 1970's," in *Toward a Better Utilization of Scientific and Engineering Talent — A Program for Action* (Washington, D. C.: Committee on Utilization of Scientific and Engineering Manpower, National Academy of Sciences, 1964), p. 76.

This frustration leads to greater exasperation when the person responsible for the projections has to guess what the policy will be. One Federal official points out that any prediction of needs for scientists and engineers is dependent on the assumptions one makes regarding the trend of activities such as defense spending. He then laments the difficulty of getting such estimates from responsible authorities:

It is impossible to get any projection of defense spending, except through the papers or when Mr. McNamara chooses to make a comment. The Defense Department, it is said, has one set it gives to the Congress, and I am sure another set it gives to the Budget Bureau. And, I am sure the analysts there have still a third set from which they are doing some of the thinking.

This last comment points up a concomitant of the situation of uncertainty with respect to manpower demands. There is a tendency for the parties involved to develop competitive strategies to forward their own interests or, in other ways, to respond in a fashion that accentuates the problem. At the same time, counter-forces arise to dampen the tendency to exaggerate the problem. Even over a fairly short period of time, a reader could be quite confused as to the *existing*, let alone the future, needs for scientific and technical manpower.

Although a serious shortage of technologically skilled manpower had been predicted for the 1960's, did such a shortage materialize? By the fall of 1964, a Congressional Committee reported:

It would appear that at this point in the mid-1960's the Nation is not suffering from a severe general shortage of trained scientists and engineers. . . . There may be a tendency to generalize from some specific or selective shortages. It would be hazardous, on the basis of such unreliable information, to make decisions which will affect future research and development work. Above all, we should be wary of leaping to a hasty conclusion that there is a crisis. It could possibly be that we are heading for a crisis. It could possibly be that we are, but we do not have sufficient or sufficiently accurate evidence to support such a conclusion.[5]

Actually a survey by the Engineering Manpower Commission had found that of the 543 companies and government agencies studied, 6 per cent fewer engineers, mathematicians, and physical

[5] *Manpower for Research and Development*, Report of the Select Committee on Government Research of the House of Representatives (Washington, D.C.: U.S. Government Printing Office, September 29, 1964), p. 69.

scientists were hired in 1963 than in 1962, and recruiting goals for 1964 were 7 per cent *lower* than those for 1963.[6]

By the beginning of 1965 cuts in defense spending were such that aerospace industry was seeking new ways of employing its skilled manpower.

The State of California, one of the most heavily dependent on defense investment, "committed itself to putting aerospace engineers to work trying to solve earth-bound problems as successfully as they have coped with space." (*New York Times,* Jan. 10, 1965) The areas in which it invited proposals were waste management, state information systems, crime and mental health, transportation systems. By November 1964, Aerojet General Corporation was already contracted to study air and water pollution on a six-month, $100,000 contract. By February 1965, similar contracts were awarded to Lockheed Missile and Space Company for a state data system, to the Aerojet subsidiary, Space-General Corporation, for a crime study, and to North American Aviation for a state-wide transportation system.[7]

It was difficult, by mid-1965, to know what the true state of affairs was. The shifts in the statements quoted appear to be in part a function of changing circumstances, of attempts to counteract overstatements in the opposite direction, and of a confusion of the part for the whole. There is no reason to believe that the prediction of a deficit of scientific and technical manpower during the 1960's was incorrect, although it would appear that the warning of an acute shortage during the mid-1960's was indeed wrong. However, the warning might have been correct if defense spending had continued to advance at the projected rate.

In Chapter 2 Bauer pointed out that what purport to be predictions are often actually cautionary tales designed to alert us to possible dangers. It is highly probable that many of the apparent contradictions in the estimates of need cited actually stem from variations in the role that different sources felt was proper for them vis-à-vis policy problems. The earlier sources may have felt it was their responsibility to *warn,* and thereby to indicate how great the deficit *might* be and make "conservative" assumptions having an upward bias. Later sources may have been concerned with the danger of an oversupply of such manpower (without any substan-

[6] Cited by Harold Wolfe in an editorial, "Today's Job Market, and Tomorrow's," *Science, 145* (August 14, 1964), 665.

[7] John McHale, "Big Business Enlists for the War on Poverty," *Transaction* (May–June 1965), 3–9.

tial intervening change in the state of the real world) and thought themselves equally "conservative" in making assumptions that could result in a low estimate of manpower demands.

Two points are crucial here. First, projections are seldom made in a policy vacuum. Hence it is important to know the pattern of opportunities and dangers that the projector envisions in deciding what his proper role is. Second, there is no such thing as an unequivocally "conservative" estimate. It is ordinarily thought that the estimate producing the lowest figures is the most conservative one, but this is conservative only if one assumes that the greater danger lies in overestimating. But underestimating demand can produce an actual future shortage. If one is primarily concerned with the availability of trained personnel, then it is "conservative" to make the highest reasonable estimate of demand. If, on the other hand, one is concerned that trained people should have jobs, it is "conservative" to make a low estimate of demand.

A First Look at Technicians

The initial survey of the manpower problem prepared for the Committee on Space in 1963 by Sharp[8] laid the groundwork for the decision to study technicians. It was clear that projections of requirements and supply were themselves dependent upon *a clearer differentiation and understanding of the nature of existing jobs, and of the relationship of education and training to the work to be done*. Additionally, the inherent instability of the projections to which we referred suggested that improvement of prediction, at least along existing lines, was perhaps not the crucial problem. Rather it seemed more important to explore the possibility of deliberate re-engineering of work roles. As a result there might emerge new roles and new patterns of relationships of education to work under the impact of technological innovation.

Obviously, any intensive study would have to concentrate on a limited aspect of the manpower problem. At that time, Mrs. Sharp suggested as one of the possible areas of research "the utilization of technicians and their role in the modern industrial structure."[9]

By mid-1964, interest in the possible expanded role of technicians in American society and the American economy had grown rapidly. The initial stimulus to this interest came from the belief that increased utilization of technicians might lead to more rational utili-

[8] Sharp, "Scientific Manpower Problems."
[9] *Ibid.*, p. 18.

zation of scientists and engineers. There had not been systematic studies to investigate to what extent this potentiality might be turned into reality. The belief that this use of technicians might compensate for the anticipated shortage of engineers and scientists was based on the European experience and on our own experience in medicine. While our own explorations indicated that these models might not be entirely applicable, this statement by two members of the National Planning Association staff represents their view on the increased use of technicians:

The expected demand for engineers would rise less sharply in the coming decade if the precedent of medicine were followed in economizing the use of highly trained professional personnel. The great strides in health in the United States in the past generation have taken place without a substantial increase in the ratio of physicians to population. Many of the routine tasks in medicine have been turned over to medical technologists and trained nurses. It is also likely that technicians could take over much of the routine testing, design implementation, and production control work currently performed by engineers. . . . In much of Europe, technical training of this type is conducted in schools offering two years of post-high school instruction in basic sciences and applied techniques. . . .[10]

Needless to say, it was predicted that there would be a shortage of technicians even under the existing pattern of usage, and from this it followed that the shortage would be even greater if there were an expansion of their role as technical support personnel. Once again, we quote a report from this period as reflecting the thinking of the time:

[During the decade 1960–1970] . . . the supply of technicians is expected to fall short of meeting the demand for these workers. The demand for technicians is projected to grow at about the same rate as the demand for scientists and engineers. On this basis, approximately 700,000 new technicians will be needed over the 1960 decade. Although information on which to base estimates of the future supply is sparse, it appears certain that, as in recent years, the demand will far exceed the supply of these workers unless technical training programs, both in educational institutions and on-the-job, are expanded.[11]

In the meantime, persons and institutions associated with the

[10] Colm and Lecht, "Scientific and Engineering Manpower," p. 74.
[11] *Scientists, Engineers, and Technicians in the 1960's — Requirements and Duties,* prepared for the National Science Foundation by the U.S. Department of Labor, Bureau of Labor Statistics (NSF 63-64, Washington, D. C.: U.S. Government Printing Office, 1964), p. 2.

educational establishment in post-high-school technical education became more interested in technician training. One author wrote in 1963

The public favors later initial employment than at high school gradua-tion. Employers favor the older employee and the one who has taken his vocational training at a post-secondary institution. The post-secondary institution will have better facilities and a more specialized staff in many fields than the high school has. For these reasons, more and more youths, when post-secondary education is available to them, are likely to post-pone their vocational training until after high school graduation.[12]

Late in 1963, the Vocational Education Bill was passed by Con-gress, providing support for technical schools and junior colleges. This was done in recognition of a widespread belief in a need for post-high-school education and training that was less than full college training.

This thinking developed through the year 1964 with the Amer-ican Association of Junior Colleges making this statement:

With the burst of knowledge and the advance of science and tech-nology, the high school must concentrate its efforts on the development of foundations that will lead to the satisfactory pursuit of further educa-tion and appropriate career objectives on the part of the majority of young Americans.

Recent examinations of this problem, the changing world of work and its educational implications, have also made amply clear the fact that a majority of the new jobs do not require four-year college programs culminating in baccalaureate degrees.[13]

It should be noted that the solution proposed in these two docu-ments — two years of education beyond high school —was not that offered by everyone. The statements, therefore, should be regarded at this point solely as reflecting awareness of an issue by groups committed to certain forms of education and training for techni-cians.

However, as we pointed out early in this chapter, by the spring of 1964 technician training was beginning to be regarded as a possible solution of the youth — especially Negro youth — unem-ployment problem. To the Committee on Space, the prospect of

[12] Harold T. Smith, *Education and Training for the World of Work, A Vocational Education Program for the State of Michigan* (W. E. Upjohn Insti-tute for Employment Research, July 1963).

[13] *A National Resource of Occupational Education,* a statement by the National Advisory Committee on the Junior College (American Association of Junior Colleges, December 1964).

exploring this possible avenue had an additional attraction. Many men of common sense saw the new technological era, the Space Age, as one in which the deeds of scientists and technologists, and their growing status in the public press if not the public mind, would lift the level of aspiration of America's youth. Yet, here and there were disquieting hints that the opposite might also be true, that less capable, less advantaged young people might quit the struggle sooner. Talks with educators indicated that in their judgment some school dropouts were the results of an increasing belief that to get a worthwhile job in the Space Age it was necessary to have at least a college degree. Hence, while the incidence of such phenomena was not known, it was, and still is to some extent, conceivable that the new technological era of which space exploration is the symbol could worsen rather than improve the plight of the underprivileged. This indeed would be irony.

We may then summarize the reasoning and assumptions that made the subject of technicians so attractive to pursue. There was reputed to be a large impending shortage of scientists and engineers. Development of technical support personnel could make the use of scientists and engineers more efficient and thereby potentially avert the supposed imminent shortage. In addition, expanded use of technicians would mean expanded career opportunities for persons who did not have the ability, means, or ambition to pursue a full college course. These persons, particularly members of underprivileged groups such as Negroes, would surely snap eagerly at this chance to get something better than was now in view for them, even though it was not a career that might satisfy a person with very high aspirations.

Needless to say, some of these assumptions proved to be less straightforward than we had anticipated. Nevertheless, this picture was sufficiently enticing to prompt further exploration.

Since the study of technicians had only recently become a matter of high interest, most of the expertise lay (and for that matter still lies) in the heads of the persons actually working on the problem. Accordingly it was planned to convene a sizable group of persons who were knowledgeable in this area.

In preparation for this meeting, three position papers were prepared. Bernard Michael, Chief of the Occupational Outlook Branch of the Bureau of Labor Statistics, who has worked with manpower problems for some years, presented a projection of future demands for technicians given the present state of affairs. He accompanied this projection with a critical evaluation of the assumptions and method employed. John Dailey, Study Director of Project Talent, a massive study of the abilities, interests, and career aspirations of children in secondary schools, explored the implications of Project Talent data for the educational preparation, career expectations, and other characteristics of future technicians. Laure Sharp, Research Associate with the Bureau of Social Science Research in Washington, and a consultant to the Committee on Space, drew on her previous research on the relationship between education and work both at the high school and college level. From this perspective, she dealt with the general topic of educational requirements for occupations, the role of socioeconomic factors in educational achievement and occupational choice, and related matters.

These papers served as the background for a conference on July 11, 1964. The participants were 21 persons from government, industry, unions, foundations, and academic institutions. Out of this conference it became apparent that certain lines of investigation warranted further explorations. So, a second conference was convened a month later, on August 11. This smaller conference consisted of 10 representatives of government, universities, foundations, and secondary education.

What follows in the next section is a revised view of the problem, based on a continued review of recently published material, the three position papers, and especially the two conferences.

Technicians: A Second Look

A main theme of this section, and of the preceding one as well, is that problems turn out to be more complicated on a second look than on the first one. However, this very complexity awakened us to some possible simplifications, so the reader should not anticipate an endless proliferation of complexity without some eventual ordering.

Who Are the Technicians?

There is far from complete agreement as to what a technician is, the sort of training needed before a person can become a technician, and the sort of persons who have the necessary ability. This lack of agreement stems in large part from the diversity of occupations embraced by the term. The following passage gives an estimate of numbers, a gross classification of the types, and the proportions of each type estimated to be in the economy in 1963.

There were roughly one million workers who could be grouped in the technician category in 1963. Of these, about one third were draftsmen and designers, two fifths were engineering and physical science technicians, and 15 percent were medical and dental technicians working with practitioners of medical, agricultural or biological sciences. The remainder served in a variety of other technician occupations; as, for example, computer programmers not classified as mathematicians. (Not included in this discussion are more than half a million nurses, most of whose jobs require between 1 and 3 years of post-high school education and training.)[14]

This listing does not do justice to the spread of abilities demanded by the range of jobs that fall under this heading. As one government expert said, "Technicians' jobs range from the very low type of test tube cleaner in the chemical laboratory up to people that I think are more professional than most dentists I have seen."

There is no consensus on what a technician *is*, since any affirmative definition will fail to include some jobs generally thought to be included either by convention or by some other definition. And this is probably no accident but rather reflects the way in which the very notion of a technician happened to arise. A technician may be defined operationally as a person whose job is less demanding than that of a full-fledged professional, and whose work ordinarily involves those portions of the technical job that can be delegated to a person of lesser training and ability. The extent of this delegation must to some degree be a function of how diligent people have been in breaking down the components of the technical task. It is ordinarily said that a technician is more than a skilled worker, implying presumably some large component of "knowledge" relative to "skill." In effect the job must be defined residually, particularly with respect to its relationship to professionals. It is certainly dependent on where one wishes to draw the line with respect to what is professional. Thus technicians are in the main people who by definition do things that require something less in the way of training and/or ability than is required of a full-fledged professional. The fact that a technician is by *definition* something less than a professional will come back to haunt us when we consider the problems of motivation associated with recruitment and job satisfaction.

[14] Bernard Michael, "Changing Requirements in the Education and Training Needed for Technician's Jobs," a document of The Committee on Space Efforts and Society of the American Academy of Arts and Sciences (Cambridge, Mass., 1964), p. 3.

What Is Their Training?

There is considerable controversy over the amount and type of training technicians ought to get. However, Michael has presented data on the actual training of a group of persons who reported themselves as technicians in the 1960 census.

The median number of years of schooling . . . was over 14 years. About two-thirds indicated some post high school work ranging from 1 year to more than a bachelor's degree. Almost half reported 1 to 3 years of college; about 20 percent reported completing 4 or more years of college and more than 10 percent indicated a bachelor's degree or more.[15]

In addition to formal training for their actual job reported by over half the persons in the study, more than three-fourths indicated that they had on-the-job training as well. Hence, the amount of training that characterizes the present technician population is quite high.

Furthermore, technician jobs as presently established require quite high levels of ability. At one of our conferences John Dailey had this to say from the evidence of the Project Talent data:

Lab technicians are right up there with college students. Electronic technicians are right up there with college students on the overall test of information. When you get into some specific tests like English achievement or linguistic interest, you find that the electronic technicians do not like verbal-type things and are bad at English. But you find in general that the technicians as such look like college kids. . . . I think we are competing directly with the colleges for our supply of highly skilled technicians, scientific and engineering aides.

This sort of data made it clear that any easy hope of training youths of mediocre capability for technician jobs had to be abandoned. It also led our conferences to the conclusion that serious consideration should be given to re-engineering some of these jobs so as to "shred-out," to use Dailey's phrase, components of the technician's job that could be handled by persons of lesser ability. This development was something of a surprise in that we had begun by considering technicians as support personnel for full professionals and had come quickly to consider the possibility of a system of support personnel for technicians — if we were to give due attention to the possibility of pulling less advanced youths into the occupational structure at a better level than they can now aspire toward.

[15] Michael, "Technician's Jobs," p. 7.

In connection with both the training and recruitment of technicians, it should be noted that the two most important sources of training for electrical, electronic, engineering, and physical science technicians (who make up the largest category — two fifths of the total in Michael's list) were technical institutes, 34 per cent, and the Armed Forces, 25 per cent. This role of the Armed Forces is important in several respects. Informed persons have commented that the aerospace industry is virtually dependent on the Armed Forces for its supply of electronic technicians. Furthermore, the Armed Forces are unparalleled by any other institution in their ability to "recruit" (usually the best word is "draft") and train youths who would ordinarily not get into a technician-training program. The Armed Forces training programs have the reputation of being adapted to the abilities of the trainees and to the specific jobs to be done. There is a minimization of the formal requirements that tend to screen out youths who might not be accepted in other programs.

How Many Are Needed?

We have already mentioned Michael's estimate that an additional 700,000 new technicians would be needed in the present decade. In a nutshell, it can be said that despite all the difficulties of estimating future demands, there evolved a substantial consensus that, under normal conditions, the demand would exceed the supply, and most of the discussion proceeded without concern for precision of the estimates. The major reason is that supply tends to create demand in the use of technicians. A government official mentioned that in studies across the country, employers regularly reported that if technicians were available they would be used. He quoted the vice-president of a chemical manufacturing company as saying:

What are you talking about chemical technicians for? They don't exist. Of course, up in Boston they are training a few of these people and the companies can get them. But in my part of the country, we can't get them. We train them ourselves. If they were available I would hire them.

An educator added that the schools have been asked to train dental technicians because dentists have just begun to discover how useful they can be:

There are dentists who don't know how to use a technician because they have never had one. But once they are accustomed to this type of person who can handle some of the operations in the ship, then they want them.

The Attractions of the Job

Project Talent surveyed high school seniors to ask them what their career expectations were.[16] The outstanding choice of the boys was to be an engineer. This preference was challenged only by the vague general category of "businessman." More interesting, when these choices were analyzed for the various levels of academic aptitude possessed by these youths, in each 10 per cent of aptitude from the least able to the most able — with one small exception — the first choice for boys at every level of aptitude was still to be an engineer. This rose from 13 per cent among the least capable 10 per cent, to 33 per cent in the most capable group. In only one of the 10 aptitude groups did so many as 1 per cent indicate they wanted to be an engineering or scientific aide. However, Dailey did point out that

The work of technicians is not necessarily unattractive to young people. When the 1960 students were asked to indicate how well they would like each of 122 occupations (disregarding salary, social standing, permanence, etc. — in fact, anything except how well they would like to do the work), technician and electronic technician in particular were relatively popular. This way well indicate that students are avoiding careers as technicians because of the lower social status as compared with careers normally associated with college or white-collar status.[17]

The picture is slightly different for girls in that from 6 to 13 per cent of the girls with various levels of academic aptitude expect to be nurses, while only in the two highest aptitude groups do a greater proportion of girls prefer to be a physician rather than a nurse. If the nurse may be considered as a "technician" vis-à-vis the doctor, it would appear that young women in our society are more likely to accept "second best" in the work arena. This impression is re-enforced by the fact that from 1 to 6 per cent of the people at each aptitude level also decided on a job as medical or dental technician. This is a fairly high percentage considering the fact that there were thirty-five jobs and the broad thirty-sixth category of "other" to choose from.

These job preferences of young women do not require additional comment. It is worth noting, however, that if one is searching for a source of technical support personnel, women are one underprivileged occupational group not to be ignored.

[16] John T. Dailey, "The Future Technician," a document of The Committee on Space Efforts and Society of the American Academy of Arts and Sciences (Cambridge, Mass., 1964), Table 1-a.

[17] *Ibid.*, p. 3.

Is Second Best Enough?

It was assumed that the job of technician would be perceived as a real opportunity for youths who could not hope to attain a full professional education. Referring to such young people, a social scientist at the conference said: "I do not think these jobs are second best. This is an opportunity for them and this is upgrading and mobility."

It would seem in retrospect that this assumption is incorrect at least in part and stems from a practice that upper-middle-class scholars have often been accused of, namely, attributing their own perception of the social structure to lower-status persons, and assuming that such persons will be "reasonable."

In point of fact, however, there seem to be a number of factors which had been overlooked. It does not follow neatly that a young Negro boy, for example, sees a technician's job as a way of finding a satisfactory niche in the Space Age job picture. In the first place, he and his school counselor may not know about such jobs. Then he might not believe that they are attainable — and he might be right. At the same time (though it would seem to be a contradiction) he might not regard "second best" as good enough. For a middle-class boy, even one of medium ability, it might be a premature admission of failure to set his sights on "second best" — even though this may realistically be the best he can achieve.

A few generalizations seem pertinent. The broadest of these is that American society is often considered to be stratified in the occupational sphere, with a sharp demarcation at the point of the job that requires a full college education. It is assumed that if one enters the structure at this point, one is regarded as "in," and there are no systematic barriers to mobility — at least if one is white and male. On the other hand, if one enters below this point, he does not have full status, nor does he have equal opportunity for mobility.

Technicians may be in a particularly difficult position. The sharp status differences between doctors and nurses have been a source of repeatedly recorded frustration and are being alleviated only by the increasing professionalism of the nurse's work in the hospital setting. A study of laboratory workers uncovered a good deal of frustration and feeling of status deprivation that seemed to stem from the fact that day-to-day contact with full-fledged professionals only made the second-rate status of the technician all the more obvious:

Is there anything in the social setting which would help to explain this result? . . . Laboratory workers, and assistants especially, typically work in close association with scientists who are of higher status in the organization and in society at large. Since the scientists symbolize the value of scientific discovery, this close association heightens interest in the work itself, and the assistants typically have such interests. But the association also constantly confronts them with a clear difference in status. All of the men are aware of such invidious distinctions as not being sent to attend meetings at organizational expense, or not being introduced to visitors in the laboratory, as well as differences in pay.[18]

This is not to say that young men and women see the situation in the work setting with anything of this clarity, but rather that they (and their parents — who may indeed have just this picture) in general draw this sort of distinction between jobs that do and do not require a college education. Interestingly, and in line with this, Dailey reports that the sons of technicians plan to become engineers.[19]

As Laure Sharp commented at one of the conferences:

Any occupation which from the outset decides that it is not a college-level occupation, but one that for admission requires less than a college degree, pegs itself at a certain level which makes it non-professional.

It is not only non-professional, but it definitely gives it the enlisted man's status which John Dailey talked about.

This crucial distinction has been observed in the system of higher education. The junior colleges have been advocated as the appropriate instrument for training technicians, and new Federal programs are geared to facilitating such "less than college, but more than high school" educational programs for technicians. The very institution of the junior college, however, has its built-in difficulty. To a predominant extent students who enter junior college view it as an avenue for later entrance to a regular college curriculum. In substance, the junior college system as it now exists serves three functions: (1) it gives two years of post-high-school training to students; (2) it gives some students an opportunity to advance to a regular college; (3) it reorients other students to accept the fact that they are not destined for a college degree or the type of job associated with such a degree.

The proposal to develop junior college curricula for technicians has considered only the first of these three functions. The proposed

[18] Stephen T. Boggs, "The Values of Laboratory Workers," *Human Organization* (Fall 1963), cited in Sharp, "Technician's Profession," p. 20.

[19] Dailey, "Future Technician," pp. 24 ff.

technician curricula, however, do not appear to offer the transferability of credits that the student who wants to go on to finish college requires, e.g., liberal arts credits may be transferable, but science, engineering, and mathematics courses tend to be considered below the par of comparable regular college courses. Furthermore, the student who enters the junior college because he does not have the requisite ability to finish college, is often not yet prepared to accept his limitations. He has entered the junior college with the intention of pursuing his education further.

Many writers, and many participants in our conferences, commented on the mystique of the full college degree that is held by middle-class families and by teachers and counselors in high schools. The educational system through the high school is presently geared to train and motivate the middle-class child for college, regardless of whether the individual child is equipped for a full college career. Acceptance of a technical program in high school for an American middle-class child of the 1960's is an early admission of failure. Hence, the junior college has the distinctive function of reorienting such youths to accept their own limitations and adjust to the fact there are other viable roles in society than those of the college-trained person. The various devices whereby this is accomplished are labeled by Clarke as "The 'Cooling-Out' Function in Higher Education."[20]

In general, then, it was the consensus of the attendees at our conferences on technicians that, under present circumstances, there was a substantial inhibition against recruitment of middle-class youths into a program for training technicians. This included even those youths who by virtue of their aptitudes might find it the best job for which they were qualified.

What About the Underprivileged?

If a less-than-college-degree job is unacceptable to youths in the main stream of the American way, perhaps it would look more attractive to someone who was outside the mainstream. But, on the other hand, what is not rewarding enough for one person may be too much of a challenge for another. Moreover, another person may not be aware of or believe in the opportunity. Or, he may decide that as long as he is going to dream he may as well dream big. All of the latter propositions are to an appreciable extent true.

[20] Burton C. Clarke, "The 'Cooling-out' Function in Higher Education" in A. H. Halsey et al., eds., Education, Economy and Society (New York: The Free Press, 1961), pp. 513–523.

When the "underprivileged" are mentioned in the 1960's, it is clear that this is usually a euphemism for "Negro," although there are occasional groups of whites equally deprived though not as systematically discriminated against. It was for this reason that at our second conference we took care to have as participants social scientists and educators who had firsthand contact with Negro youths.

There was a constant reaffirmation of themes, which have since worked their way rather regularly into the press:

1. Negro children are at a disadvantage by the time they enter grade school. Many of the verbal skills that white children pick up in the family environment are not transmitted to the Negro children.

2. Even the best of the tests of ability are not culture free. Negro children tend to be systematically underrated by such tests.

3. The school system re-enforces the disadvantage. Teachers who know the handicaps under which the Negro child labors unwittingly treat them in such a way as to slow up their development — quite the opposite from helping them to overcome their slow start.

Circumstances such as these led Dailey to warn in the first conference that we be under no illusions about the possibility of training for technicians "those disadvantaged who have been doubly screened on selective service and very, very thoroughly and negatively screened on some pretty good trainability tests." There was a strong consensus among informed conferees that for very large portions of Negro youth it is illusory to hold out technicians' jobs — even redesigned and simplified technicians' jobs — as a viable occupational future. The damage had occurred too early and was too difficult to reverse. What is needed is an early program of cultural enrichment. (The nation has since inaugurated Project Head Start.) More appropriate tests for assessment of ability could also help. (The American Psychological Association and many individual psychologists are directing their efforts in this direction.) And, the school system must be restructured to help rather than hinder the Negro child. (There are now various "enrichment" programs, but it is not clear to us whether all these are appropriate.)

What does all this have to do with the supply of technicians? Again, it is an example, like that of projecting the need for technicians, of a problem which evaporated for practical purposes. It will be remembered that debate over the need for more technicians dissolved when good evidence was introduced that supply creates demand in the use of technicians. Similarly, once we looked at the issue of how the prospect of a career as a technician could keep Negro drop-outs in school, we encountered a social problem of such magnitude that an attempt to aid it by dangling a single job category in front of Negro youths would be obviously fatuous. Surely, it would be ludicrous to advocate aiding Negro youth in developing their abilities merely in order to ensure a supply of technicians. The broader moral and social issues are so compelling that our initial concern is dwarfed to obscurity.

In a word, opening up technicians' jobs for Negro youths can be of some help, but it is not going to solve the problem of school drop-outs among Negroes. For a very large proportion of them, technicians' jobs would be beyond their competence.

Other problems were uncovered that are more narrowly specific to potential technicians' careers. It has been a generation or so since most American families looked to their teen-age members as a potential source of income. Hence it is easy to overlook the fact that the income of many Negro families is so low that there is pressure on teen-agers to leave school and get a job. Earlier training programs which aimed at getting drop-outs back into some educational activity tended to include a legal stipulation that the student could not earn additional money. However, the urgency of permitting the student to contribute to the family's income has been recognized in the Economic Opportunity Act, and students will be permitted to earn several hundred dollars a year. In the words of a Federal official involved with this program, this provision "is intended to help keep them in school until they can complete their training."

The job of technician was reported to be invisible not only to the Negro student, but also to his counselors. Here are the results of a study of Negro high school graduates in Washington, D. C., reported by an educator:

Apparently the guidance people have very little real information available to them, either as to the capabilities and interests of the children or the real nature of the labor market. They advised most of the children who said they wanted to go to college that the fields that were open were in the physical sciences, the result being that a large proportion did not make out too well.

This same educator reported flatly that, in general, neither the students nor the counselors were aware that most technicians' jobs existed. Upon reflection, the reason became apparent. A child learns about occupations in two basic ways. He may learn from direct experience: he knows someone who does the work, or he is able to see someone doing the work (e.g., policeman or teacher). Or, he may learn indirectly via the mass media. The technicians' jobs are generally invisible by either route. They are not jobs performed in settings where lower-class children are likely to observe them, and they are of sufficiently high status that the child is unlikely to know someone holding such a job. On the other hand, technicians' jobs are not sufficiently glamorous to get much attention in the media. News about the space program, for example, features astronauts, scientists, and engineers, but not technicians. A conference participant put it this way:

The higher-status occupations are described in the mass media and in more or less formal places. You can get plenty of writing about engineers and scientists. The lower-status jobs, the ones toward the bottom, are generally familiar. Some of them have been around for a long time and a lot of them are sufficiently numerous that they would become known to counselors on a sheer chance basis, while the middle-range occupations have neither the visibility via print that the high-status ones do, nor the general socially shared knowledge of them.

We have examined a child's limitations in learning about job possibilities. Why the counselor does not know about them is less reasonable and less explicable.

The importance of the reported "invisibility" of technicians' jobs is perhaps most strongly underscored by the apparent exception, that of medical technician. A Negro educator pointed out:

That is where you will find the greatest number of Negro kids who have gone into technology. They have seen the hospital. They go to the labs as field experiences for science classes.

This same educator has inaugurated a summer placement program for high school students in settings where they will be in contact with jobs of middle-range status. As an initial step this appears to be a successful one.

While we have earlier discarded the idea that technicians' jobs can *solve* the problem of Negro drop-outs, we do not in any sense want to understate the importance of the invisibility of middle-status occupations to the Negro student. Realistic striving is dependent upon the perception of realistic opportunities. Competent authorities at our conferences confirmed the reality of our concern that the Space Age might pose special difficulties for the child who has his opportunities limited. It is clear that without a perception of a career ladder that is realistic, many Negro children alternate between fantasy and despair. The following is an interchange between several conference members:

First social scientist: The question is whether this (publicity for the space program) has increased their perception of being able to achieve more, or whether this publicity for the space program and for science has on the contrary caused them to feel more hopeless about their future.

Educator: The youngster whose perception has increased is the really bright youngster who is going to achieve in spite of handicaps. So he begins to see new doors opening for him. But for the large mass of kids, it seems to me that they are in really greater despair.

Government official: Any flights of fantasy about becoming astronauts?

Second social scientist: Functionally, despair and fantasy are probably just as destructive with respect to the problems we are talking about.

Educator (with considerable emphasis): And they go very much together.

Hence, the existence of middle-status occupations may have an importance beyond their sheer physical numbers. If they are made visible, *and* if presently underprivileged youths are convinced that they are accessible (there are many, many stories about Negro youths being skeptical on this score, and for good historical reason), they can create realistic aspirations rather than the destructive alternatives of fantasy or despair.

With this statement of the importance of middle-status occupations, it may be appropriate to turn to a reassessment of the overall problems of technicians in the Space Age.

9

New Perspective
on Technicians

(Written with Laure Sharp)

Rapoport has characterized the Space Age as the age of "technocratization." He says

The technocratization process seems to involve a rational planning dimension, a more fluid set of arrangements among institutions and professional roles, and a greater detachment from historical and local concerns in the direction of a more universal and cosmopolitan orientation.[1]

This is a statement of *one set of forces* that has been activated in the Space Age, and it is that set of forces that is most in harmony with the new technology. But, it does not exhaust the important trends and situations in the society that are relevant to our moving into the Space Age with ease and comfort.

For example, Rapoport states quite correctly that we are coming increasingly to judge people by what they can do rather than by their background. (The importance of family background was considered at some length in the preceding chapter.) This is true; but what they can do is generally a function of their education and training, and their education and training are a function of their background. Therefore, it is more correct to say that *if a person can overcome the handicaps of his background,* we are increasingly ready to judge him by what he can do regardless of his background. Neither deliberate discrimination, nor formal requirements, nor financial restrictions are needed to effect a *de facto* disadvantage for a child pursuing education. All that is necessary is that his parents have been poorly educated; this in turn automatically

[1] Robert N. Rapoport, ed., "NASA and the Technological Revolution," a document of The Committee on Space of the American Academy of Arts and Sciences (Boston 1965), p. 60.

144

makes more probable a set of conditions — euphemistically called "a deprived cultural background" — which places the child at a disadvantage before he even enters school.

Historically, mobility in American society was kept open to a large extent by the multiplicity of routes whereby one could rise in status. Certainly, among the present generation of leaders in business, government, and unions there are many who had less than a college education. This is obviously less true among leaders in education, science, and engineering — all spheres of perform-ance highly dependent on formal education. Indeed, the Space Age already appears to be an era in which success in most of the more rewarding and more important jobs in society will be in-creasingly dependent on formal education. While we are moving into an age in which equality of opportunity will continue to be a value, we are at the same time moving into an age in which mental competence is also increasingly valued. If some of the social dy-namics which have come to our attention are allowed to proceed unimpeded, the Space Age *could* become a caste society of the edu-cated and the uneducated in which the children of the uneducated would find it more and more difficult to escape the caste of their origin.

It is clear that we have no intention of letting such a state of affairs develop. This would both violate our values and deprive us of much valuable potential manpower. However, a consideration of this issue from the point of view of the specific issues and prob-lems arising from the present state of technicians' jobs may suggest some of the particular steps that should be taken.

Flexibility and Inflexibility

While we have not introduced the issue of "flexibility" into the discussion up to this point, it is one of the concepts that comes up most frequently in the discussion of the use of all levels of man-power in the rapidly changing technological scene. But there would be no need to stress the desirability of flexibility except that there already exist elements of inflexibility. Furthermore, there is an in-creasing trend to introduce devices of inflexibility which has prob-ably not been fully realized. Virtually all movements to "raise standards" introduce elements of inflexibility — though this need not necessarily be so.

The easiest and most usual way to raise job standards is to in-crease the educational requirement. In fact, the achievement of a certain educational level usually acts as a validation device for being considered for a particular level of job. The fact that a person has a bachelor's degree does not guarantee that he can do the job nor that someone without the bachelor's degree could not do it. It is assurance to a prospective employer that the candidate has a certain level of intellectual competence, a certain degree of socialization, and a certain level of knowledge — possibly more, possibly less of each than is required. In plain words, it makes the selection process simpler.

In a similar fashion, one occupation after another is moving toward "professionalization," toward setting up standards based on specific training and experience that make *horizontal* mobility, even for persons with a high level of education, increasingly difficult.

The introduction of the devices for reducing "flexibility" are customarily justified by the increased demands of the job and increased competence and standards of the persons holding those jobs. There can be no doubt but that this justification is in varying degrees unwarranted, and sometimes totally warranted.

We have seen, however, and will look even more explicitly at, the difficulties that can be created by the generation of requirements that are, or that are perceived to be, inflexible in connection with the technicians' jobs. At this point we want to raise the general issue of the care with which formal requirements for access to jobs will be handled in the future.

Formal requirements are generally regarded as criteria of competence to perform. They also may be a license to perform after one has lost competence; or a barrier to one who is competent to perform but does not have the formal qualifications. The clear implication, in our view, is that we shall probably have, in many areas, to move progressively away from formal criteria of evaluation to direct measures of ability to perform — a task not easily achieved even after many decades of industrial psychology.

The Educational Barrier

The technician-level job poses a dilemma. It demands too little education to satisfy one group and too much education to be attainable for another group. True, between these two extremes there may well be enough persons, including women, who have the ability to get the education but not the ambition to reject the job, to fill all of the technicians' slots that may exist in the future. However, our analysis convinced us that the problem of filling technicians' jobs was a minor one compared with the associated problems that unfolded. Furthermore, solution of the associated problems would automatically solve that of recruiting and training technicians.

Somewhat earlier we mentioned the inherent difficulty posed by the fact that a technician's job is defined as less than a professional job. This is compounded by the increasing equating of access to professional work with at least a bachelor's degree. Historically, it has been possible for technicians to become "non-degree engineers." Even in the civil service, technicians make use of the opportunity to study, take exams, and rise to the status of engineer. And, there is no reason to believe that in the future persons who have become technicians in the past, e.g., blue-collar workers who studied part time, graduates of technical institutes, or former members of the Armed Forces, will not find the boundary between professional and technical support status somewhat permeable.

However, in an era in which the full college degree is seen increasingly as the key to full-status jobs, a boundary job such as that of technician poses a difficult educational choice for the youth who is not ready to relinquish his ambitions. Ironically, one of the complaints of persons who have to make estimates of manpower supply and demand is that the line between technician and engineer is so poorly defined. They point to the fact that many "engineers" are really technicians in terms of the work they do and the abilities they possess, while many "technicians" do more demanding work and have more ability than many engineers. The clearer definition of the status of technicians might make estimates of supply and demand easier and discussions of the job more precise. But, it may be this very impreciseness that is the salvation of the present situation.

The dangers of developing a caste structure cannot be alleviated solely by keeping open the *possibility* that a technician may rise to full professional status. What must be avoided is the creation of a sharply defined cleavage line in the occupational structure that is clearly linked to a college-noncollege cleavage line in the educational structure.

As it becomes more evident that the job content of all technical jobs is in a constant state of flux, it is realized that formal education must train not only for the entering job, but also for the capacity to be retrained for new jobs. This has led to more emphasis on teaching basic scientific disciplines to engineers and, more recently, to the desirability of so doing with technician candidates.

If this is done to the point that the youth entering a junior college will have full transferability of credits in the event he can and wants to continue for a full college degree, then the drastic decision of making the college-noncollege choice on leaving high school can be avoided. In the event that he does not have the capacity to go on to college, the "cooling out" procedures of the junior college can reorient him to accepting a less than professional beginning job.

The work-study, or "cooperative," formula of combining higher education with a job has existed for some decades without receiving wide acceptance. Yet, it currently has strong supporters in the educational establishment, is supported by recent government programs, and is reported by NASA officials as a highly desirable procedure for training NASA technical personnel. It has the advantage of making it possible for youths of limited financial means to begin and/or complete higher education. It has the further advantage of deferring the college-noncollege decision for the youth of marginal college ability. He can test his academic ability while simultaneously getting started on a career line that can either be extended to full professional status, or curtailed to a more limited subprofessional one.

With respect to the work-study program it is easier to specify what the content of such programs should be than to recruit capable youths into them. In the past and at the present, work-study plans have had the appearance of a distinctly second-rate educational route. It is not easy to see how in the immediate future the middle-class youth with full professional ambitions will deliberately decide on the work-study program of higher education. (Soviet efforts to make this route mandatory and universal, by the way, produced completely farcical results. Appropriate jobs could not be found for *all* secondary school students, and bright students succeeded in beating the system.) However, it may prove an acceptable route for middle-class boys and girls who cannot get into the other educational channels.

Having considered the situation of those youths who are reluctant to aim at a job that is less than of full professional status, we now turn to the other group, the underprivileged. The work-study approach may be a highly acceptable route for the underprivileged youths for whom the necessity of beginning to earn money at an early age is urgent, and who have had little if any preparation for higher education. Here society faces a major task to which it is beginning to give attention. Such programs as Project Head Start at least reflect our recent realization of how early in life the damage sets in, the fact that many of the abilities that make for success in later life are acquired before school age, and that it is very difficult to offset this disadvantage at later periods. There is no reason for us to stress this matter, since it is now getting a good deal of attention — begging the question of whether the new programs will in fact be successful.

There is an issue, however, which is only beginning to be perceived. If we may take the Negro youth as our example again, it is not sufficient to give them the opportunity to acquire the necessary abilities. It is also necessary to convince them that the acquisition of these abilities is worthwhile, that they will be a usable commodity.

There are two related problems in reorienting the Negroes' perception of the occupational structure. Like all lower-status persons they have a selective perception of the occupational structure. We have seen how middle-status occupations can be socially invisible to this group since they are not observed directly or treated with any fullness in the mass media. Clearly one task with which we are faced is that of communicating a fuller picture of the range of job opportunities, particularly those in the middle range, which represent an improvement over the poorer jobs that Negroes see as accessible and are more attainable than the upper-status jobs that they may see as beyond their reach.

An obvious route for conveying information about middle-status jobs is via counselors. We have heard few, if any, flattering statements about the effectiveness of counselors in guiding the careers of lower-class children. Perhaps the existing counseling system is overloaded. We make no pretense of judging this nor of prescribing a specific solution. Many possible avenues are obvious. We only point to the fact that the *existence* of these jobs must be conveyed if we are to expect lower-status boys and girls to aspire to them.

The second point is that Negroes must be convinced that these jobs are indeed accessible to *them*. The casual reader of the newspapers may assume that the passage of the Civil Rights legislation and the making of opportunities available to Negroes will produce an automatic rush to take advantage of these opportunities. The Negro's experience has been too long and its effects too deep seated. There are militant Negroes who fight for the rights of their race. But even among the militants there is a tendency toward deep skepticism when opportunities are apparently presented to them. For example, Negro college graduates are presumably members of the Negro community who have ability and initiative and are looking for opportunity. However, when a group of employers organized a meeting at New York's Statler-Hilton in June of 1965, they were befuddled to find they had to convince these young Negroes that they were serious in their offer of job opportunities. "It can take a long time for a company to get across the message that it's really prepared to hire Negro graduates on an equal basis."[2]

In the same week, *The New York Times* carried a story that openings in Project Head Start were going unfilled:

A director of one of the agencies said: "The Board of Education people just sat in the schools and expected the parents to come."[3]

This experience can be repeated for many programs designed to create equal opportunities. And presumably the pattern holds for groups other than the Negroes. Side by side in the *Times* with the story about the vacancies in Project Head Start was a story about the use of a resident of Appalachia to convince his neighbors that the government's poverty program was a sincere effort.

It is not sufficient to develop abilities and motivation among the underprivileged: we need to give them an appropriate perception of the world in which they may realize their ambitions. Ability, motivation, and perception of the world go hand in hand, with no one element necessarily leading to others. It will be essential that an active effort be made to go out and bring these people into the program. The picture of the evolving society is even more demanding than that of the present, and, unless the various elements of the poverty program are successful, the underprivileged group may drop out of school at an accelerated rate.

Somewhat more novel is the picture that emerged at the boundary between the technician's job and that of the scientist and the engineer. Most of the elements of this picture were not new, but they came together to give a sharper image of a potential cleavage point in the occupational structure. Perhaps the technician's role is uniquely designed to illustrate the nature of this cleavage point, since it is explicitly defined as *not* professional, yet it demands most of the skills and values of the professional. Moreover, since it places the job incumbent in daily contact with the person he aspires to be, all these strains are accentuated. This sharply drawn

2 "Industry Plugs Positive Negro Job Message," *Business Week*, No. 1869 (June 26, 1965), 34.
3 "Operation Head Start Reports 3,000 Vacancies for Children," *The New York Times*, July 8, 1965, p. 15.

picture, while it may almost be a caricature, could alert us to similar situations arising at other points in the structure where formal requirements differentiate adjacent jobs, whether vertically or horizontally. It is frustrating not to be able to move up, but it may be almost as frustrating not to be able to take a job for which one is equipped at the same level. The urgency of this consideration lies not only in the satisfaction of the individuals involved, but in the necessity for the "flexibility" that is so much touted as being essential to a rapidly changing work situation.

In closing, we feel we owe a debt to the problem discussed in Chapter 8, that of manpower planning. We have quoted a number of people on the difficulty of predicting both the demand and the supply of people for various jobs. And we have indicated that they have convinced us of that difficulty. We have also indicated that when a number of problems were looked at closely, precision of prediction did not seem as important as it had seemed initially. If we look seriously at the future confronting the next generation, or the generation not yet in school, all these observations become more and more pertinent. The major task will not be to train people for specific beginning jobs, but rather to decide what level of education we can achieve for the population as a whole, and what mechanisms can be established for a continuous process of retraining and re-educating throughout one's work life.

We do not, of course, anticipate that everyone will want or be able to attain the same level of competence. But, both from the point of view of the opportunities we owe to our people, and from the point of view of the demands on competence that are likely to be made on the work force, it would be preferable to have an over-educated and overtrained work force who, if necessary, could turn their abilities to finding constructive new tasks on which to work.

In the past, we have been concerned as much with an oversupply of trained people who could not find jobs, as with a shortage of people to fill vital functions. If our economy is to be as productive as it promises to be, our preferred strategy should be to make sure that we will meet the minimum requirements of abilities that the economy will need in the foreseeable future. To this extent planning is in order. Concern for oversupply should become a secondary consideration, with the assumption that such a surplus will be absorbed either by the development of new useful roles, as just suggested, or by increased leisure.

This cannot, of course, be done immediately. The difficulties of our immediate task of keeping lower-status youths — again, males in particular[4] — as effective members of the work force is such that we are far from the time when we will have to worry about what to do with their excess abilities.

Another dimension of the problem is with us now, however. That is the problem of continuing training and education.[5] The problem of the rapid obsolescence of scientists and engineers has become familiar. A parallel need for adult education of executives has also been recognized in the past few decades. Yet we have barely begun to grapple with this question. In general, we have assumed that "old dogs can't learn new tricks," and we have done little to study how adults learn, and more particularly how they can be taught to learn continuously as they go about their daily affairs. An immense amount of adult learning and training is taking place, but we have studied the learning habits only of rats, pigeons, and children.

Nor have we considered systematically where the responsibility for retraining lies, or, more pertinently, who can be made to assume this responsibility, the setting in which it should take place, and so on. These are matters to which our immediate attention should be directed.

---

[4] This statement is not a bow to the decline of male dominance. It is a reflection of the superior job opportunities of lower-class *women* in service industries, for example. And in the case of Negroes, it reflects the existence of some established lines of mobility for educated Negro women, particularly as teachers and social workers.

[5] Various aspects of the problem, as well as specific suggestions for mechanisms, financing, responsibility, and so on, are found throughout Laure Sharp's paper, "Technicians in the Space Age," and Rapoport's "NASA and the Technological Revolution."

PART **III**

The Process:
The Way Effects Happen

10

The Transfer
of Space Technology

(*Written with Richard Rosenbloom*)

A new device merely opens a door; it does not compel one to enter. The acceptance or rejection of an invention, or the extent to which its implications are realized if it is accepted, depends quite as much upon the condition of a society, and upon the imagination of its leaders, as upon the nature of the technological item itself.[1]

We have been in a process of convergence. The book began with a statement of broad considerations for studying second-order consequences of technological developments. It moved to a general consideration of some actual or alleged effects on public opinion, communities, occupational choice, our imagination, and the like. Then, one imputed effect, on manpower, and especially on the demand for and supply of technicians, was considered in detail. Our approach was somewhat that of the sociology of knowledge in that the topic was treated in terms of the way in which it evolved as we studied it.

As we pursued this path of convergence we have considered in a relatively sketchy fashion, from time to time, *how* these effects take place. Here, in discussing technology transfer, we focus directly on one example of the process whereby a large technological enterprise has its impact on society. Technology transfer can be defined as the process by which the use of technology is diffused throughout society. In the case of the space program, the results of technology transfer are the much-discussed technological by-products — the spill-over — of space technology. Yet transfer is, in

[1] Lynn White, Jr., *Medieval Technology and Social Change* (New York: Oxford University Press, 1962), p. 28, commenting on the Anglo-Saxon failure to realize the full potential of the stirrup.

155

essence, a social process, and here we will look at the structure of that process.

Among the second-order consequences of the space program the technological by-products present special opportunities for study. Whereas, as we have repeatedly noted, many secondary effects of large-scale technical change are unintended, unanticipated, and often unwanted, the realization of technological by-products of the space program is intended, expected, and definitely wanted. NASA has acted to focus public attention on the opportunities for technology transfer, has predicted the realization of significant benefits, and has established special programs to facilitate that outcome.

Technological by-products represent only one of several economic effects of America's space efforts. In the early years of this decade, there was considerable discussion of the possible adverse macroeconomic effects of what some described as a "diversion" of scarce technical resources from the support of innovation in the private sector. A recent period of relatively slow economic growth, coinciding with the remarkably rapid growth of military (and then NASA) expenditures for R & D on increasingly exotic systems, lent weight to the arguments of skeptics.[2] A more flourishing economy in recent years, however, has quieted that concern. Another category of economic effects arises through the direct application of space technology to important needs. The practical application of satellites — as, for example, in systems for communication, weather forecasting, and navigation — has demonstrated the value, here on earth, of our growing extraterrestrial capabilities.

Our concern, however, is not with either the macroeconomic consequences of space efforts, or with the direct applications of space technology, although both are interesting and important. Facilitating the realization of indirect benefits — through utilization of the technological by-products of space efforts — constitutes the focus of interest in technology transfer.

---

[2] For a statement of the argument that military and space R & D were not "spilling over" into the private sector, see Robert A. Solo, "Gearing Military R & D to Economic Growth," *Harvard Business Review, 40,* No. 6 (November–December 1962), 49–60.

As examples of the concern for this question in the popular press, see: "Is the Moon Race Hurting Science?" *The New York Times Magazine,* May 26, 1963; "Is the U.S. Moon Project a Waste of Brain Power?" *U. S. News & World Report* (July 20, 1964); Amitai Etzioni, *The Moon Doggle* (Garden City, N.Y.: Doubleday & Company, Inc., 1964); and Edwin Diamond, *The Rise and Fall of the Space Age* (Garden City, N.Y.: Doubleday & Company, Inc., 1964).

The framework of our analysis is the same as that postulated in the Introduction, to guide the work of the Commitee on Space. In this case, of course, the problem is not one of an unwanted excess of secondary effects, but of a lack of anticipated benefits. New products and processes for everyday life, as the phrase often appeared, "should" be forthcoming from the rapidly accumulating accomplishments of space technology. High expectations encouraged the search for evidence of success and the development of means to broaden and accelerate the natural forces at work. Hence it became necessary to define the *products* that might be expected and the process that would produce them.

In studying this subject we followed the basic strategy adopted for the work of the Committee on Space, namely:

. . . to improve our ability to handle problems such as this, rather than either (1) speculate on the full scope of the issue, or (2) concentrate on one or a few substantive studies. . . .[3]

During the period of study there was no shortage of material speculating on the subject. Systematic studies of some of the substantive issues were more rare, but several projects were initiated at that time.[4]

The two sorts of activities, of course, are interrelated and both have an impact on the evolution of policy and practice in a new field of enterprise. Speculation, debate, even polemic (and this subject elicited all three in abundance) serve to identify the salient issues and to direct the focus of policy-oriented substantive inquiry.

[3] See the Introduction to this book, p. 3.
[4] Most noteworthy among the speculative works are Donald N. Michael, *Proposed Studies on the Implications of Peaceful Space Activities for Human Affairs* (Washington, D. C.: U.S. Government Printing Office, 1961), pp. 113–124; "The Implications of Technological By-Products," a 1962 Summer Study at Iowa State University, sponsored by NAS/NRC; and the NASA Conference on Space, Science, and Urban Life, Oakland, California, March 1963. The only sizeable substantive study completed at that time was a NASA-sponsored survey, *The Commercial Application of Missile/Space Technology,* Denver Research Institute, Denver, Colorado, September 1963. Among the studies completed more recently are Sumner Myers, "Technology Transfer and Industrial Innovation," *Looking Ahead, 15,* No. 1 (February 1967), 1; Robert Solo, "Gearing Military R & D to Economic Growth," *Harvard Business Review, 40,* No. 6 (November–December 1962), 49; Donald Marquis (Director), *Annual Reports,* M.I.T. Research Program on Management Science and Technology; Edward B. Roberts and Herbert A. Wainer, "Technology Transfer and Entrepreneurial Success," *Proceedings,* Twentieth National Conference on the Administration of Research, October 26–28, 1966, pp. 81–92; and the study of channels of technology acquisition by Gilmore and others at the Denver Research Institute.

The interaction of these, speculative and substantive, shapes the perceptions of policy makers and can guide their actions. How this happened in the early development of the Technology Utilization Program at NASA, and how questions of this sort might be approached, in general, constitute the coordinate axes of our inquiry.

The Contemporary Discourse

Prevailing conceptions of the nature of the products to be expected, the process which might produce them, and the timing and scale of their occurrence provided the basic elements for formulation of a program to deal with the problem. As the following review of what can be called the "contemporary discourse" will show, the prevailing expectation was

1. The potentially useful technical by-products of space efforts were of substantial magnitude in aggregate.
2. These by-products were realized in the application of tangible, identifiable developments to meet specific needs outside the space program.
3. The "normal" process by which opportunities for use of new technology were identified and applications developed and implemented was inadequate on two counts: important possibilities might be overlooked and those which were exploited would take too long.

Let us begin with the first of these points. There was widespread acceptance, in the first years of the space program, that the impressive stream of technical developments directed towards needs of the program would also contain a number of innovations that could find fruitful application in everyday affairs.[5] In the earliest days of the nation's space activities, and despite the otherworldly character of missile and space technology, a starry-eyed citizenry heard promise of wondrous benefits here on earth. Even the best informed and most sober commentators were enthusiastic. In December 1958, with the ink barely dry on the Space Act, General James Gavin offered a catalogue of "space-related" developments *already achieved* which ranged from high-speed computers to Pyroceram.[6]

---

[5] Concern for this subject was enhanced by apprehensions about adverse macroeconomic and scientific effects of the space program, as noted earlier.

[6] General James Gavin, address to the International Bankers Association, Bal Harbor, Fla., December 1958.

On the third anniversary of the passage of the Space Act, in October 1961, the Congress published a study of the practical values of the space program, more than half of which was devoted to discussion of its "economic values" and "values for everyday living." Its purpose was

. . . to explain to the taxpayer just why so many of his dollars are going into the American effort to explore space, and to indicate what he can expect in return which is of value to him.[7]

In discussing benefits for everyday living, it stated

The so-called side effects of the space exploration program are showing a remarkable ability to produce innovations which, in turn, improve the quality of everyday work and everyday living throughout the United States.[8]

The report then devoted three pages to an extended excerpt from Gavin's catalogue of "space-related" developments and added that "hundreds of other examples of the space program's value for everyday living could be cited."[9] The following are some of the examples of the transfer of space technology cited in the report:

A plasma arc torch [has been] developed for fabricating ultrahard materials and coatings by mass production methods.

Medical research, and our health problems can use such things as film resistance thermometers. Electronic equipment capable of measuring low-level electrical signals is being adapted to measure body temperature and blood flow. In a dramatic breakthrough, illustrating the unexpected benefits of research, it has been found that a derivative of hydrazine, developed as a liquid missile propellant, is useful in treating certain mental illnesses and tuberculosis.

Ground-to-air missiles that ride a beam to their targets must measure the distance to the target plane with an accuracy of a few feet in several miles. This principle, now being applied to surveying techniques, has revolutionized the surveying industry.

Silicones for motor insulation and subzero lubricants — used in new glassmaking techniques for myriad products.

Automatic gun cameras — used in banks, toll booths, etc.

Fluxless aluminum soldering — used for kitchen-utensil repair, gutters, flashings, antennas, electrical joints, auto repairing, farm machinery, etc.

---

[7] *The Practical Values of Space Exploration*, a report of the House Committee on Science and Astronautics, U.S. Congress (Washington, D. C.: Government Printing Office, October 1961), p. 1.

[8] *Ibid.*, p. 43.

[9] *Ibid.*, p. 47.

Inert thread scaling compound — used by pump manufacturers serving process industries.

Satellite scan devices — used in infrared appliances, e.g., lamps, roasters, switches, ovens.

Automatic control components — used as proximity switches, plugs, valves, cylinders; other components already are an integral part of industrial conveyor systems.

Missile accelerometers, torquemeters, strain gage equipment — used in auto crash tests, motor testing, shipbuilding and bridge construction.

Space recording equipment automatically stopped and started by sound of voice — used widely as conference recorder.

Optimism about the technological and economic benefits of space technology was not limited to the Congress. In March 1962, Lyndon Johnson described the national expenditures on space as "investments which will yield dividends to our lives, our business, our professions, many times greater than the initial costs."[10] Implicit in discussion of the magnitude of the transfer "problem" (really, an opportunity) was this prevailing conception of the nature of the product and the process of technology transfer.

The overwhelming scale of the mountain of rapidly accumulating technical information published by Federal agencies and their contractors gave rise to concern that important needs might unnecessarily remain unmet because of an inability to identify new possibilities buried in an avalanche of publications. For example, Congressman Roman C. Pucinski, Chairman of a House Subcommittee concerned with methods for handling information said, in 1964,

I am convinced somewhere in the mountainous scientific literature there is a cure for cancer, a solution for harnessing limitless energy from nuclear fusion and a way to solve the food requirement of a growing world population.[11]

The head of the Small Business Administration told another Congressional committee that the Federal government had

. . . an obligation to develop a workable system of utilitizing this enormous reservoir of scientific information so that its benefits can be transmitted to businesses both large and small in order to provide the ingredients necessary for accelerated growth in our civilian economy.[12]

[10] Remarks of March 16, 1962, in *Congressional Record*, Vol. 108, No. 57 (April 12, 1962), A 2857–2858.

[11] *Washington Daily News*, April 21, 1964.

[12] U.S. Congress, House of Representatives, Select Committee on Government Research, *Hearings* (Washington, D. C.: U.S. Government Printing Office, 1964), p. 741.

In this way the opportunity to exploit by-product applications of space technology was translated into a responsibility as well. As James Webb, NASA Administrator, told the Oakland Conference on Space, Science, and Urban Life, in March 1963,

We in the Space Agency do not seek to justify our program on the basis of the industrial applications which will flow from it . . . since we are committed to this great effort in space, however, a responsibility exists to glean from it the maximum public benefits which can be obtained.

Even severe critics of the space program could applaud acceptance of this responsibility. Amitai Etzioni, for example, in his polemic attack on the program, *The Moon Doggle*, argued for a reduction of 80 to 90 per cent in the NASA budget and called the program "an economic drag, not a propellant."[13] In virtually the same breath, however, he concluded that, whatever the level of spending on space goals, "we must look for ways to increase the economic fall-out of Federal research."[14]

NASA Response

The administration chose to take positive steps to meet the responsibility for technology transfer recognized by Mr. Webb. As Dr. George L. Simpson, Jr., then Assistant Administrator for Technology Utilization and Policy Planning, told the Congress in the 1964 NASA Authorization hearings:

NASA is committed to a hard-driving effort to transfer the useful fruits of our research and development effort to the private sector of the economy in as quick and useful a way as possible.

These prevailing ideas about the nature of the problem, the character of the product, and the mechanism of the process necessarily shaped the conception of the NASA program for technology transfer.

The *products* were expected to be discrete, countable innovations in which a specific technical advance within the space program was identified as relevant to a need outside the program, adapted, and applied.[15] If these, indeed, were the products, then the *process*

[13] Etzioni, *Moon Doggle,* p. 72.
[14] *Ibid.,* p. 80.
[15] The listing of by-products on pp. 159–160 illustrates the tendency to phrase this phenomenon as a simple matching of space-related achievement and non-space need.

could be considered in several parts, namely, identifying potentially useful technical developments, disseminating knowledge of those developments to firms which might adopt them, and promoting the adaptation of the technology so that it would be directly useful in the new field.[16]

To the extent that the *problem* of technology transfer was considered to be a problem of retrieving useful technology from a vast "reservoir" of innovations buried in a "mountain" of publications, it could be met by methods of screening, identifying, evaluating, and disseminating information about *specific* technical achievements. This, in fact, became the core of the NASA program. Dr. Hugh Dryden described it as follows, in an address to a management audience in July 1964:

NASA collects, organizes, and disseminates scientific and technical information on its in-house and contractor work by means of the latest in computer, microfilm, and other methods. Decentralization with local access is the keynote.

To be of most use this information must be identified and repackaged to meet the special needs of the non-aerospace industries. This is the purpose of the Technology Utilization Program, the first national program of its type. Staffs at each NASA Center are systematically organizing the reporting of new or improved ideas, materials, methods, and technology developed in the course of research and development activities. The more promising reports go for analysis to a group of industrial research institutes familiar with current industrial needs and requirements. If the analysis indicates promise, reports are published and indexed so that they are readily available to any segment of the economy that may find them useful.

A NASA document, dated April 15, 1963, describing the Technology Utilization Program, stated that the objective of the newly-designated Office of Technology Utilization was "to disseminate this new knowledge to industry quickly, in the form of ideas and innovations, which can be translated by industrialists into new products, processes, techniques, devices, and materials for the benefit of the Nation's consumers." An exposition of the procedures established by the Office begins with the statement:

The primary problem, of course, is the identification and retrieval of these ideas and innovations from both NASA's Field Centers and contractors.

---

[16] The last step proved to be a crucial stumbling block. Matching new technology to an existing need would not automatically bring about the use of the technology. Of this, more later.

If "transfers" were specific, identifiable, countable, it made sense to identify and to count them. The first substantive study of significant scale related to the transfer of space technology tried to do just that. It was commissioned by NASA and carried out by the Denver Research Institute in 1963. Its aim was:

. . . to identify tangible economic by-products of missile/space programs which have or are expected to find commercial use, and to determine the origin, and circumstances surrounding the origin, of these by-products.[17]

Following the pattern established two years earlier in the House report, NASA officials commonly spoke of the number of transfers accomplished, and the number of innovations "retrieved." For example, Louis Fong, then Director of the Technology Utilization Program, was quoted in *Steel* as saying:

A recent study by Denver Research Institute, Denver, uncovered 185 examples [of spin-off] and added: "These examples do not include all . . . of the transfers which have taken place." . . . NASA centers have come up with about 850 innovations thought to have industrial potential.[18]

Difficulties Emerge

The simple paradigm outlined on the preceding pages led to difficulties. Early enthusiasms, bolstered by lists of "transfers," later gave weight to the arguments of skeptics, who soon were able to point to the lack of substantial new results. In February 1964, for example, *Space-Aeronautics* joked that the program designed for "fall-out" and "spin-off" had resulted in "foul-up" and "spin-in." Their headline asked: "NASA Claims 1,000 Spin-offs. Where are They?"

The report of the study by the Denver Research Institute actually disclosed that it had not proven possible to find a significant number of specific by-products. More important, the authors suggested that the term, itself, was misleading in that it "implies that the contribution comes in the form of direct and readily identifiable results to a particular effort, when, in fact, most contribution does not." Two of the principals in that study, writing later in the *Harvard Business Review*, concluded that the nature and magnitude of the potential

---

[17] *The Commercial Application of Missile/Space Technology* (Denver, Colorado: Denver Research Institute, September 1963), p. vii.
[18] "Hitchhiking on Space Technology," *Steel, 153* (December 1963), 26.

benefits would have to remain speculative. "To argue now that this benefit is negligible, as some have, or tremendous, as others have, is future . . ."[19] they said.

At the root of these difficulties lay several crucial and unanswered questions about the nature of the transfer process. Our expectations about the way in which things naturally proceed will necessarily shape our efforts to influence the course of events. NASA's program to accelerate the utilization of the technological by-products of its primary activities rested on a simple model of the process. As the inadequacies of that model became more apparent, several consequences ensued.

First, the lack of an adequate understanding of the process was — and still is — the main obstacle to formulation of an effective policy for technology transfer. As Welles and Waterman observed, during the early days of NASA's Program,

There is wide disagreement in government circles about how to enhance diffusion of technology, one reason being that no one really knows how technology gets diffused.[20]

Thus, it became apparent that the nature of the process had to be explored while steps were taken to affect it. An experimental approach was needed. As a recent comprehensive summary of issues and practices in this field put it:

At the early stages of the technology utilization program there was a tendency for NASA to highlight specific products or techniques of the "spin-off" type. Critics were quick to point out that these instances were rather rare. . . . In more recent years, NASA has described the activity as an experimental program to find better methods of transferring technology.[21]

But the adoption of an experimental approach in this case raises as many issues as it answers. Blind experimentation can be wasteful or even harmful. If experimentalism is to be enlightened, rather than blind, it too must be guided by some model of the structure of the process which new policies seek to facilitate.

In fact, although knowledge of the transfer process is incomplete,

[19] J. G. Welles and R. H. Waterman, Jr., "Space Technology: Pay-off from Spin-off," *Harvard Business Review* (July–August 1964), 106.

[20] *Ibid.*, p. 117.

[21] *Policy Planning for Technology Transfer,* a Report of the Subcommittee on Science and Technology to the Select Committee on Small Business, United States Senate, prepared by the Science Policy Research Division, Legislative Reference Service, Library of Congress (Washington, D. C.: U.S. Government Printing Office, 1967), p. 22.

it is not nonexistent. An assessment of our present knowledge can serve to suggest "amendments" to the simple model, yielding conclusions that could guide the design of experiments. The merits of the conclusions could then be tested against empirical results.

Two principal amendments emerging from such an analysis, however, tend to throw doubt on the ability of NASA — or any other institution in comparable circumstances — to cope effectively with the requirements of a program for transfer of the technical byproducts of its primary work.

The first of these amendments recognizes that by-product effects probably will not ensue primarily in the form of specific innovations. The second recognizes that the process is, ordinarily, a multistage sequence, carrying beyond the identification, repackaging, and dissemination of information to include activities concerned with the further development of the technology and the implementation of the by-product innovation.

In the next section we shall examine the nature of the transfer process, with the aim of substantiating the validity of these amendments and their central importance to an understanding of the process as a whole. In the concluding section, we shall return to the idea of an experimental program and consider, in the light of our expanded conception of the nature of the process, the evident limitations on the competence of a large mission-oriented agency to explore or influence that process.

The Process of Transfer

Technology comprises the practical media by which man interacts purposefully with his environment to meet his needs. Its components may be tangible or intangible; technology includes "hard" tools, such as the plow, the drill press, and the airplane, and "soft" tools, such as production scheduling procedures, double-entry bookkeeping, and computer programs. In either case, the essence of technology is cognitive, not material; a tool, whether simple or complex, whether Bushman's bow or modern man's missile, is the embodiment of man's understanding of the natural world. Each expresses some man's grasp of an essential principle and society's ability to preserve and transmit that insight through time and across distance. Hence, in common usage the term technology may refer to either its cognitive state — the aggregation of ideas, understanding, and systematic ways of doing things — or its operational state, the embodiment of those ideas in usable things and procedures.

As I said in the beginning of this chapter, technology transfer is the process by which the use of technology is diffused throughout the society. Although it usually implies information transfer, technology transfer goes beyond that. It involves, in addition, the reduction of that information to practice in a new operational setting. Transfer may be as simple as convincing farmer A to use a seed used by farmer B. More often, in a modern industrial society, it implies a complex intermixture of scientific, inventive, and innovative activities. The generation of new knowledge, and the recognition of previously unseen practical implications of knowledge, new or old, are all common constituents of this process.

As Harvey Brooks has suggested,[22] there are two dimensions to technology transfer. Along one dimension technology moves from the most general to the most specific. This involves a transition from purely cognitive to operational states, from principle to practice, as from Hertz' law to Marconi's wireless or from the Hahn-Strassman discovery to Fermi's atomic pile. This may be thought of as the "vertical" dimension of transfer.

The use of a particular technical element also may spread "horizontally," as when it is transferred from use in one context to use in another. The most simple form of horizontal transfer occurs in the diffusion of relatively standard products through a market comprising numerous potential users, as in the introduction of consumer appliances (home laundries, television, and so on) and innovations in producers' goods (diesel locomotives, digital computers, plastic containers). Horizontal transfer may also lead to new uses largely different from existing ones, as, for example, in the first applications of controlled nuclear fission to commercial power generation, of electronic computers for routine business data processing, or of P.P.B. techniques outside the Defense Department. In such cases imitation will not suffice, and the technology must be embodied in new forms, which serve to adapt it to the different context. The task of adaptation may require the discovery of new knowledge, and the invention of new devices or techniques, as well as the development of a form suited to the new environment. Once it is adapted for initial use in the new context, the technology may be imitated by adoption in other comparable settings.

Technology transfer is a pervasive aspect of technical work in an

[22] Harvey Brooks, "National Science Policy and Technology Transfer," in *Proceedings of a Conference on Technology Transfer and Innovation, May 15–17, 1966* (Washington, D. C.: National Science Foundation, NSF 67-5, 1967), p. 54.

advanced society. The nature of our social institutions affects the flow of technology horizontally and vertically. For example, the science-based, mission-oriented organization, as exemplified by NASA in the public sector, or Am T&T in the private sphere, is a "social invention," which has greatly facilitated the vertical transfer of technology. Organizations of this type make heavy investments in technical work intended to support and advance the organization's extensive operational activities, and they organize themselves in ways intended to facilitate the translation of more fundamental discoveries into operational technology.[23]

The structure of firms and industries also affects horizontal transfer. In the United States, commercially aggressive independent firms in producers' goods industries, such as machine tools and plastics, commonly serve the needs of a heterogeneous set of customers. The supplier firm often acts as a channel for transfer between diverse fields. Machine-tool innovations, for example, such as improved cutting tools or electronic control systems, were developed to meet the special needs of particular industries, such as automotive and aircraft, but are adapted and vigorously promoted to a wide spectrum of users through the normal commercial activities of the industry which serves them in common.

Analogous institutions and practices perform this function within science. Alvin Weinberg has observed[24] that as scientific fields grow a hierarchy of theory and application emerges. The common body of theory provides a link between increasingly diverse fields of application, and is a means of exploiting developments in one field throughout a wider domain. The institutions of technical education also are a means for vertical transfer. Discoveries made on the frontiers of science are "packed down," in Derek Price's phrase, and gradually incorporated into the common body of knowledge on which professional education is based.

Information transfer bears on the process of technology transfer in three substantial respects: (1) the interchange of information about new technical possibilities and about operational needs leads to identification of opportunities for innovation, and provides the basis for the dissemination of information to create widespread awareness of those opportunities; (2) the dissemination of information about new technical possibilities is used to develop and extend

---

[23] See, for example, Jack Morton, "From Research to Technology," *International Science and Technology* (May 1964), 82–92.

[24] Alvin Weinberg, *Reflections on Big Science* (Cambridge, Mass.: The M.I.T. Press, 1967), Chapter 2, "Scientific Communication," p. 39.

the competence of technical professionals and provides the basis on which they carry out the technical work needed in the development and implementation of innovations; and (3) the accumulated body of technical information in a field is drawn upon by scientists and engineers for specific problem-solving purposes in the same tasks.

The specialization of technical work by field, and the structure of industrial organizations, while in some respects facilitating information exchange, also act to impede certain kinds of horizontal information transfer. Any individual or organization is open to only a limited number of sources of technical information. The information-seeking behavior of professionals follows reasonably consistent patterns and, for most engineers and scientists in industry, interpersonal communication structured along work or professional relationships accounts for a substantial fraction of the incidents of information exchange.[25]

The relevance of a new element of technology follows no such predictable pattern. Knowledge gained in oil-drilling operations and new developments in welding techniques may both relate to the science of metallurgy and contribute to the development of general knowledge in that field or benefit from new extensions of that knowledge. Thus information discussed in an article in the *Oil and Gas Journal* or at a meeting of the American Petroleum Institute might ultimately find application in the work of a welding-process engineer in a company making heavy electrical machinery, and hence find its way into the pages of *Design News* or the program of a meeting of the Society of Welding Engineers.

The medium of information transfer appropriate for any particular match between new technology and operational need is likely to be as unpredictable as the match itself. The information may be explicit, as described in papers, drawings, and so forth, or it may be embodied in the devices themselves. Both are important means by which new capabilities are diffused through the society. Also of great importance, however, is the implicit transmission of information (pointed out earlier in this book) as it becomes incorporated

---

[25] The work of the Committee on Space included several inquiries into the information-seeking practices of engineers and scientists in industry. See, for example, C. P. McLaughlin, R. S. Rosenbloom, and F. W. Wolek, "Technology Transfer and the Flow of Technical Information in a Large Industrial Corporation" (Cambridge, Mass., 1965). Rosenbloom and Wolek subsequently extended these studies (originally conducted at the Harvard Business School) under grant from The National Science Foundation. See their "Technology, Information, and Organization," Division of Research, Harvard University Graduate School of Business Administration, currently being printed.

in the general competence of an individual or working group and is brought to bear in a new context as the work of that individual or group changes over time. The tasks of an organization may change in focus, the expertise of a group may be drawn on by others using informal channels of communication, or individuals may move to new jobs, perhaps in different firms or different sectors of the society. In fact, scientific entrepreneurship is an important means by which the capabilities of skilled individuals are brought to bear on new types of problems. It may well be one of the most significant ways in which technology developed in one sector finds application in another.

Studies of the Transfer Process

The "natural" process of transfer and diffusion obviously is an imperfect one. The implications of a new technical development may long go unrecognized, and even when they are perceived by someone, there may be no awareness of the opportunity on the part of those in a position to take effective action, or an opportunity may be recognized but left unexploited because an organization does not feel able to do so.

Scholars in several disciplines have long been concerned with the study of innovation by imitation in a variety of fields, including education, agriculture, medicine, and industry.[26] Everett Rogers offers a general framework for analyzing the diffusion of innovation by imitation as a sequential process which includes, from the point of view of the adopter, creation of awareness, investigation, trial, and then commitment to use. Sociological and economic variables have proved to be productive factors explaining the rate and penetration of adoption of new technology in widely varying circumstances. One of the major lessons of this tradition of research is that those who would promote the adoption of innovation must take into account the varying circumstances of the context of the prospective adopter, as well as the possibilities inherent in the technology itself.

[26] The following are typical of the traditions of research on the diffusion of innovation: J. Coleman, K. Katz, and H. Menzel, "The Diffusion of an Innovation Among Physicians," *Sociometry* (December 1957); Z. Griliches, "Hybrid Corn: An Exploration in the Economics of Technological Change," *Econometrica*, 25 (October 1957); Elihu Katz, "Notes on the Unit of Adoption in Diffusion Research," *Sociological Inquiry*, 32 (1962), 3; Edwin Mansfield, "A Model of the Imitation Process," *IRE Transactions on Engineering Management*, 9 (1962), 46–50.

An excellent survey of more than 500 studies is given in Everett Rogers, *Diffusion of Innovations* (New York: The Free Press of Glencoe, 1962).

Granting that imitative diffusion of innovation is a simpler process than that likely to be involved in the transfer of space technology, the design of a program to facilitate its working would pose a substantial burden for any single agency that chose to undertake it.

Findings on the imitative adoption of innovation, furthermore, may have only limited relevance to a program for the transfer of space technology. For the most part, these studies are concerned with situations in which an individual is the adopter, e.g., a farmer for hybrid corn or a doctor for new pharmaceuticals. The space program must be concerned necessarily with promoting adoption by organizations, primarily business firms, rather than individual users. Furthermore, few technical innovations originating in the space program can be expected to be applicable to everyday needs by simple process of imitation.

Space technology must usually be adapted before it can be adopted for other use. The new needs to be served are linked to the original use by analogy, rather than by more direct similarities. As was discussed earlier in this book, analogy implies the identification of structural or functional identities in otherwise dissimilar situations or things. Transfer occurs by analogy when someone perceives a similarity between characteristics of a discovery or invention and some aspects of a need or opportunity in another situation.[27]

If secondary innovation must come by analogy, rather than by imitation of the original innovation, there is a new element in the process — creative adaptation — and hence a different kind of problem. The practical significance of this difference can be great; not only has a step been added, but it is one calling for the exercise of originality. How is this adaptation to be conceived, and by whom: the original innovator, the potential adopter, or some third party? The addition of this step probably will increase the lead time from original invention to spread of the derivative innovation. The new conception of a fruitful adaptation will probably have to go through some process of development and refinement before it is suitable for implementation. Finally, it seems likely that potential adopters will not be identified until after adaptation of the source invention. In contrast, where we are concerned only with imitation of innovation, definition of the innovation usually implies an identification of the

---

[27] In one sense analogy is at the root even of imitation of innovation. The farmer considering trial of the hybrid corn being used by his neighbor must first come to believe that, with respect to those factors relevant to the use of the new seed, his situation is largely similar to his neighbor's. While the analogy in that case may seem obvious, objectively similar situations are not always so perceived by involved participants.

potential users, e.g., doctors for new ethical drugs, farmers for agricultural developments, such as hybrid corn or diesel locomotives.

What then, is implied by a commitment to transfer "the useful fruits" of space R & D to serve other needs? As a recent Congressional report phrased it:

The underlying theme of NASA's Technology Utilization Program is, put simply: New technology has no value until it is used, and it cannot be used by an organization unaware of its existence.[28]

It is equally clear that awareness, while necessary, is not a sufficient condition for use. Granted, the exchange of information is a requisite condition for the transfer of new technology, whether it be from person to person, firm to firm, industry to industry, or government to private enterprise. The transfer of information can create awareness of opportunities for innovation and also contribute to the technical competence required to exploit those opportunities in particular situations. These are necessary steps; they set the stage for transfer, but the success of the play depends upon the behavior of the principal actor, the adopter. The decision to proceed with an innovation and the effective action necessary to see that it is successfully implemented are equally necessary and equally hazardous elements of the transfer process.

Consider the problem of promoting the introduction of new products based on space technology. In the *Harvard Business Review* survey of businessmen's attitudes on the space program in 1963, 74 per cent indicated the belief that a payoff from the space program through "new products for our everyday lives" was almost certain or very likely.[29] There are a number of new technical fields from which one might reasonably expect that useful new products would emerge. Instruments and electrical components, control systems, compact power sources, new materials, bio-medical devices, telemetry, and new fabrication techniques are among those often mentioned.

Having achieved "awareness of the existence and potentialities" of a new development in one of these fields, one may still be far from having brought about its use in other settings. These first, and necessary, steps must be followed by others, more difficult for a

[28] *Values and Benefits of Space Exploration,* report of House Subcommittee on NASA Oversight (Washington, D. C.: U.S. Government Printing Office, 1967), p. 47.
[29] E. E. Furash, Jr., "Businessmen View the Space Effort," *Harvard Business Review, 41* (September–October 1963), p. 14.

governmental agency to influence. The technology applied for use in the space program will probably have to be developed to meet the quite different objectives and constraints of its new usage — a process calling for the investment of funds at some risk. Furthermore, once developed, its introduction as a new product must meet the criteria applied by firms in making such decisions — adding a further set of hurdles.

Clearly, the secondary use of technology designed for space applications must impose criteria of performance that can only have been of secondary importance in the shaping of the new technology for its original purposes. The civilian application must be adapted to its different purposes by the process of development. In the process the trade-offs between characteristics desired for civilian use are necessarily different from those of the space program. For example, the design for a satellite solar cell will call for maximum reliability, at a necessarily high initial cost. That trade-off would be reversed in the development of civilian applications for these devices. As another example, discussion of medical benefits derived from investments in the space program in a recent Congressional report calls attention to techniques for remote monitoring of a patient's condition. The report points out, however, that

It should be emphasized at this time that there are differences between aerospace applications and more conventional applications of these measurement techniques. The major needs which characterize the aerospace application of measurement techniques are those which are remote, continuous, automated, and of relatively long duration. The flight requirements also necessitate that the measuring equipment be of high reliability, small size weight, and have sufficient power requirements so as to operate in an environment of high acceleration and vibration, without severely restricting or causing discomfort to the subject.[30]

Differences of this sort are what account for the cost, delay, and uncertainty, which result from the need to develop the original innovation to meet new circumstances.

There are, of course, a number of cases in which firms have invested successfully in the development of new products based on space technology. In the early years of the Technology Utilization Program, for example, the press reported the following "transfers":

The Fairchild Camera and Instrument Corporation demonstrated a prototype of a video tape recorder for home use which was iden-

[30] *Values and Benefits of Space Exploration,* p. 42.

tified as an off-shoot of work by a subsidiary of the company in instrumentation tape recorders.

The Bell Aerosystems Company developed a patented electronic stethoscope, which can eliminate all unwanted heart sounds and bring into focus only those that the physicians want to hear.

The Barnes Engineering Company developed a micrometer to measure the diameter of hot steel rods in the course of production, using infrared components developed originally for horizon sensors in satellites.

These examples illustrate secondary applications of inventions to fill analogous needs in consumer, professional, and industrial markets. The difficulty of promoting utilization of new products, however, is illustrated by the fact that only one of these three derivative inventions was put to use by the firm which developed it. The news report on the Fairchild prototype acknowledged that "the machine does not fall into Fairchild's present marketing pattern and the company hopes that some other concern will adopt and produce it." Bell Aerosystems is primarily in space work and although it had tested the stethoscope in clinics, hospitals, and schools, it offered manufacturing rights to medical electronics companies. "It's not in our line," said the company spokesman. In contrast, infrared technology was the principal line of the Barnes Engineering Company, which was proceeding to develop and market the special micrometer.

Promotion of technological by-products as marketable products must take into account the procedures by which companies commonly evaluate new products before introducing an innovation. The nature of the decision process, as well as such oft-cited characteristics as "creativity" or "innovativeness" are factors to be reckoned with. Product development activities in most firms are inbred to a large extent, and the most important reference groups for innovation and product ideas are internal. In appraisal of new products, the idea of a "thread of continuity" is an important theme. Thus studies of the criteria applied in evaluation of product ideas list such factors as interrelations with existing product lines; common use of materials, facilities, distribution channels, or existing markets; peripheral benefits to existing products; and use of the firm's distinctive know-how. Competitive pressures, or a strong champion for a new idea in an influential position in the company, can ease the way for the new product idea, but in their absence, novelty may not flourish.

The difficulty of promoting the adoption of specific innovations arising from the space program has led some to argue that general technology and "intangible" results offer the greatest opportunity. The design of space systems and devices have been optimized for very specialized and complex functions which seem unlikely to be economically adaptable to other needs. More promising opportunities probably lie in new methods, new design approaches, and new production techniques. This is the conclusion, also, of the researchers at the Denver Research Institute, who argue that "intangible spin-off is far more important than tangible spin-off."[31]

These indirect and intangible benefits are represented by the many advances in manufacturing techniques, electronic components, the processing of materials, small portable energy sources, and so on, which can be expected to be incorporated as incremental improvements in manufacturing processes and in the design of industrial and consumer products. Advances of this sort, in general technology, are important for two reasons. Individually, these incremental improvements make possible gains in the efficiency of industrial operations or advances in the quality of performance of new or established products. Although the gain from application of a new welding technique may be small, the aggregate benefits of many such advances, applied in many industries and firms, can be quite large. Furthermore, the convergence of a number of such improvements, along with technical advances arising in other fields, may make possible new fundamental inventions of substantial individual significance. The realization of most important inventions in the past has been contingent upon developments in materials, energy sources, or other aspects of the technological competence of the society as a whole. Although the role of these minor technical improvements — which can flow from the massive Federal R & D programs at a staggering rate — may be difficult to trace to specific civilian applications or to associate with important new products or processes, their potential contribution must be large.

As is true for direct applications, realization of the potential inherent in this general technology is neither simple nor immediate; these incremental benefits usually will need to be adapted to the differing requirements of the new field for which they are relevant. In many instances the identification of relevance, and the process of development and adaptation, may have to be carried out through a new institutional context. A different sort of organizational innova-

31 Welles and Waterman, "Space Technology," p. 108.

tion will be required to take advantage of the host of potential by-product applications of this sort.

Conclusion

Technology transfer is the process by which practical knowledge is acquired, developed, and put to use in a context other than the one in which it originated. In order to assess programs, policies, and prospects for the transfer of space technology, we have been concerned with three main aspects of the subject: the nature of the product, the structure of the process, and the characteristics of the recipient systems. Simply put, we have tried to ask: what can be transferred, by what means, and to whom?

The products of technology transfer are operational innovations introduced in some other sector of the society. They may take the form of new or improved goods and services, production processes, or changed practices in any purposeful endeavor. Space technology may lead to these innovations through the direct adaptation of some specific device or procedure, or through the indirect contributions of "intangible" advances in general technology. The opportunities for transfer of general technology seem more substantial than those for specific innovations, and the adaptation of space technology by analogy seems more probable than the opportunity for diffusion by imitation.

The process of transfer comprises several different kinds of functions. If innovation is to occur, information about new technology which now resides in the minds of engineers and scientists working under Federal contract, or in the countless pages of technical reports, must somehow be retrieved, transformed, and introduced to use. In most cases, the development and implementation of the innovation takes place within the firm. Thus, it is the private sector of the economy, i.e., firms and industries, that constitute the main host system for technology transfer.

Information transfer is one aspect of the transfer process. It creates the necessary articulation between the body of space technology (in its cognitive state) and the population of potential users. Information may be transferred in various forms — explicitly, as in documents, or implicitly, as embodied in certain products or in the persons of skilled professionals. Thus the movement of people and products, as well as of papers, leads to the diffusion of technical information. Different means of creating the requisite articulation for information flow are called for in different sectors of industry;

a mechanism that is effective with science-based firms probably will not be so with others, one that is appropriate to very large firms may be less so for small ones.

However, much more than information transfer is involved. Another requisite function for technology transfer is the transformation of the technology and the identification of its implications for use. As we have argued, the creative adaptation of space technology may be an essential ingredient in any program attempting to couple that technology to other uses. Furthermore, the acquisition of knowledge and recognition of its implications must be followed by constructive action to implement an innovation. Often a new development must find a champion who will bring about the necessary sequence of coordinated actions to produce the solid substance of innovation from the shadowy stuff of knowledge.

Some Implications

The chief elements of the NASA Technology Utilization Program deal with the information transfer components of the process. They seek to identify, repackage, and disseminate information about advances in space technology. Whatever assumption one makes about the main characteristics of the opportunities for the transfer of space technology, an information transfer strategy raises difficult issues.

To the extent that the space program will be yielding technology in the form of specific innovations that might be imitated or readily adapted for other use — and this appears to be a small part of the opportunity for transfer — a program geared to identifying the implications of those innovations and disseminating information about them can facilitate the creation of adopter awareness which is the necessary first step in transfer. But, as we have argued above, the disposition to act is as significant in innovation as awareness of the possibility for action. Unfortunately, the former is much more difficult for an outside agent to promote. The use of new developments, either as new products or major process changes, is contingent upon decisions within a firm to invest in the adaptation and introduction of the technology.

The dilemma which thus arises is as follows: Is it better for an agency to concentrate on half a loaf, emphasizing information dissemination and avoiding involvement in the more difficult subsequent stages of innovation, or to accept responsibility for doing something about the entire process and therefore also become involved in areas fraught with practical and political difficulties?

In the alternative case, where transfer takes place through indirect contributions to the "general" technology of industry, a comparable issue arises concerning communication links with organizations and the internal functioning of those organizations. Here, of course, communication is less concerned with information about specific innovations than with disseminating "state of the art" and other sorts of information, which enhance the effective level of technical competence in an organization. An effective strategy for this purpose probably should pay more attention to information-flow processes, the transfer of skilled people (rather than documents), and other mechanisms that tend to bear on general competence rather than specific problems. The needs of target organizations, however, vary quite substantially in this regard, depending upon their size and whether or not they are technically sophisticated (i.e., science-based). In its most ambitious form this sort of goal becomes one of raising the level of technical competence in all industries and, as such, is unmanageably large.

Obviously, much more is involved than is implied by the technology itself. The "host" systems must be understood and new programs at the federal and regional level have to be coupled to relevant processes within those systems. In the United States, the target system for technology transfer is the firm; decision-making processes for development and innovation and methods of maintaining general competence in the technical organization are likely to have an important effect, therefore, on the rate, character, and effectiveness of technology transfer.

Some Broader Questions

A Federal program to facilitate the utilization of space technology is, in effect, a form of intervention in the "natural" system of technology transfer at work in the society. We have tried to analyze the functions in that system in order to anticipate how it might respond to the special problems posed by space technology. To analyze a system, one tries to "take it apart," separate out functional elements, and understand how they may be made more effective. Often, however, it is the interaction of these elements, rather than their separability, that is crucial to the over-all performance of the system. A "systems point of view" must deal with its "wholeness" as well as its differentiable parts.

When a large institution attempts to affect the external system through which the second-order consequences of its actions are worked out, it encounters two common dilemmas. Both can be discussed here in terms of the question of technology transfer, but both have a wider relevance as well.

The first dilemma arises out of the conflict between the restricted functional capabilities of the institution concerned and the broad functional requisites of a program that would affect the system as a whole. In the case of technology transfer, NASA's ability to develop programs whose function is to alter the character or the accessibility of the technology itself is much greater than its ability to develop programs that would alter the receptivity to technology of potential innovators in firms. Yet failure to do the latter may reduce the effectiveness, in terms of the goals of this program, of the former.

The significance of this may be grasped in terms of a simple analogy. In designing a space system, the agency takes care to establish its control over all of the elements of the system. It uses that control to ensure that the characteristics of subsystems are appropriate to the needs of the system as a whole. Yet this essential characteristic of its behavior in respect to its primary mission cannot be carried over into efforts to deal with the external systems that determine the nature of the second-order effects of the agency's programs. Hence, trying to cope with these second-order efforts is a little bit like trying to design a space system in which the agency develops a vehicle to be launched by a booster of unspecified design to be produced through an informal collaboration of unidentified organizations.

The second dilemma results from the fact that we don't really know how these social systems function. Thus, even when it becomes possible, indirectly or otherwise, to influence a greater range of functions bearing on the process of technology transfer, the agency cannot be sure that it has reached the relevant ones or that it understands how they fit together. To pursue our previous analogy, one might establish control over both the booster and the space vehicle only to discover at a very late date that the available grade of fuel necessary to operate the booster is inadequate for the task at hand.

Thus to say that NASA may not be able to exert a substantial influence on the transfer of space technology is not an indictment of the Agency, but rather a general observation about the limited range of competence of any large institution. It is set up to deal with its mission, not to manage the second-order consequences thereof. Yet the society must deal with these consequences, and a first task is to learn how that might be done.

If the job is defined as one of "learning how" rather than of producing immediate results, both sorts of limitations can be met. A learning model for the development of a technology utilization program, in the first place, is the most appropriate response to the absence of sound understanding of how the system works. At the same time, such an approach can address the problems arising out of limited functional capability. Jerome Wiesner[32] nicely characterizes the nature of such a system as follows:

> The essential properties of a good learning system for a society can be simply stated. First, the individual experiments should be small enough so that a single failure will not endanger the society, and also it should be possible to carry out experiments rapidly. In general, these two requirements mean many separate experiments going on simultaneously. Second, the communication channels should be arranged to provide information about performance quickly. Third, the error detection system should be sufficiently sensitive to detect malfunction early enough to allow adjustments to be made before the detrimental effects are too damaging. An essential part of the error detecting process is clearly defined objectives or goals, for without standards of feedback process cannot function.

Is a learning model a viable administrative concept for a mission-oriented Federal agency? It implies a commitment to goals rather than to any particular set of means. It demands a form of experimentalism and readiness to discard ineffective approaches that may well be incompatible with the nature of such an agency and its staff.

Even if the agency could create conditions in which it would be practical to be responsive to feedback about performance, the nature of the appropriate feedback is extremely difficult to specify in this case. For example, as we have shown, early discussions of the performance of the Space Technology Utilization Program concentrated on direct measures of tangible results, on documented "transfers," innovations actually spun-off to industrial use. Unfortunately, measures of that sort are extremely difficult to develop, since a program for technology utilization, like a research and development laboratory itself, is concerned with the creative transformation of

---

[32] Jerome Wiesner, *The Challenge of Technology*, Proceedings of the National Industrial Conference Board, Annual Conference on Science and the Humanities (New York, Nov. 30, 1966), pp. 6–7.

knowledge. The ultimate payoff is dependent upon a chain of events going beyond the program itself and involving the capacity to effect innovation as well as the ability to generate the technical basis for it. Tangible measures, furthermore, would be an incomplete measure of the contributions of such a program. Other functions are important in an over-all assessment of technology utilization, including creation of awareness of new possibilities and a greater disposition to act on new technology.

The answer might lie in a program for technology transfer located at the Federal level in an agency not directly concerned with primary missions like space or defense. This would permit a broader view of the body of technical knowledge available as a result of government R & D, and might reduce the commitment to demonstrating the success of experimental programs, as opposed to experimenting for the sake of learning about the transfer process.

The stated goal of NASA's Technology Utilization Program is "to shorten the time gap between the development of new knowledge and its broad and effective utilization."[33] This transcends the primary mission of the agency and constitutes a broad social goal. If the society is seriously to pursue that goal, it must first embrace the intervening goals of developing knowledge and institutional competences adequate to the task.

[33] Testimony by Breen Kerr, Assistant Administrator of NASA, before the Subcommittee on Advance Research and Technology, Committee on Science and Astronautics, U.S. House of Representatives, April 1, 1966.

Conclusion

The Space Program and Social Consequences Reconsidered

In the broadest sense, NASA and the space program are contributing to a general, important trend in American thinking and ways of doing things. The distinction between the private and public domain in the economy and political life is becoming blurred. A decade ago, Kenneth Galbraith's cry for a greater emphasis on public goods and services seemed if not a radical objective, at least one that would be difficult to attain. Yet, the shift has been taking place without much fanfare, and with a surprising amount of acceptance on the part of American business. It has become almost trite to say that "private enterprise" should become involved in the "crisis of the cities," in manpower training, community development, and housing, and in such other public problems as air and water pollution, transportation, and the like.

In the meantime, there is a growing awareness of the necessity to establish criteria of performance for the evaluation of a wide range of activities which cannot be measured against the traditional yardsticks of the market place. Some of these activities involve public programs, some involve private production of public goods and services, as well as research and development in the private sector.

In other words, there is a merging of the private and the public sectors, and of market and nonmarket criteria of evaluation, with a concern for evaluating more than the direct output of individual enterprises and programs, and the thinking reflected in this volume is much more prevalent than such thinking was six years ago when the American Academy project began.

Research on Complex Social Phenomena

While the avowed goal of the social sciences is to develop knowledge and methods for guiding social affairs, there is a considerable lag in the rate at which social scientists turn their attention to problems we consider "really important." The difficulty lies in the fact that the "important" complex problems of the world seldom come packaged in ways familiar to social scientists, and seldom are reducible to dimensions or forms that the social scientist is accustomed to handling. When important and complex problems have been of interest to social scientists for some period of time they gradually develop ways of relating to them. For example, problems of war, peace, and related phenomena have now been discussed for some time in *The Journal of Conflict Resolution*. A tradition of research on problems of conflict has developed. However, only *aspects* of such large problems as war and peace can be researched in systematic fashion.

And, this must be so. Large, complex social events and problems are not *reducible* to systematic terms, though they have systematic aspects. The space program, the troubles of our cities, the war in Vietnam, all reflect some systematic processes that we find in other contexts. But they also have their idiosyncratic components and their fortuitous ones. To understand, predict, or control the course of events, one must effect a synthesis based on systematic knowledge and on the idiosyncratic features of that situation with which he is confronted.

Our emphasis on devices of anticipation, the modern approach to planning, social indicators, and so on, reflects this circumstance. However well we develop our systematic knowledge of the ways in which large technological enterprises affect society, that knowledge will have to be applied in specific circumstances. We need to understand those specific circumstances, hence the need for measures of what is, and devices for anticipating what may be.

Straight factual research, such as that on manpower needs of the space program, the state of the communities in the Cape Kennedy area, and so on, is of practical operational use. However, except as it may contribute to more general systematic understanding of the *sorts* of phenomena involved, such research will be of limited value. Our goal, as iterated and reiterated in this volume, was to move toward the development of much more systematic knowledge.

When a large, complex entity such as the space program comes onto the scene, research related to it can be organized only gradually because the conduct and funding of research are themselves part of social systems which must change in some perceptible form in order for such research to take place. Most social research is done in orthodox academic departments by scholars geared to researching traditional problems on which they are judged by their peers, and as a result of which they receive academic promotions. Furthermore, research financing comes from institutions that have developed their own programs and procedures. These have a built-in inertia not easy to change. One can say that the social systems of research and funding have a bias against novelty. It is much easier to get money and to attract researchers to a project similar to what is already being done.

In general the starting up of research in such a new area as the study of the social impact of space exploration is a matter of patient matching up of current world problems with traditional problems of research. In our own efforts we concentrated on the matching of substantive issues of the social impact of space exploration to traditional problems and areas of research. The historians, sociologists, psychologists, and the like whom we persuaded to join us in our work were scholars trained in the traditions of their disciplines. We were able to stimulate them by offering support from an institution, the American Academy of Arts and Sciences, which lay outside the regular academic structure.

The result, as the reader by now well knows, was the study of *aspects* of the social impact of space exploration and, it will be recalled from earlier chapters, we touched on only a few of the aspects that might have been studied.

In discussing our substantive findings we have laid strong emphasis on the concept of the "host system." We have said repeatedly that the second-order consequences of the space program are and will be as much a function of the social system on which the space program impinges as they are a function of the space program itself. The same model must hold for attempts to develop a tradition of social research. It cannot be accomplished by brute force, but must depend in large part on one's ability to activate the dynamics of the ongoing research community, to exert initiatives that will blend in with the stream of events.

There are two instances in which success has been apparent. By a happy set of circumstances our work on *Social Indicators*[1] seems to have contributed momentum to national concern with that issue. Similarly the work of Rosenbloom and his colleagues on technology transfer blended in with and contributed to the development of that area of study.

[1] Raymond A. Bauer, ed., *Social Indicators* (Cambridge, Mass.: The M.I.T. Press, 1966).

However, it cannot be expected that the success or failure of our efforts to contribute to the over-all problems of understanding and managing the second-order consequences of large technological enterprises will be documentable. Rosenbloom has pointed out that technology transfer is a complex process in which the impact may so blend with other events that its identity becomes lost. The same, regrettably, may be true for the impact of our own efforts.

The Intricacy of Second-Order Consequences

The bulk of previous research on second-order consequences of technological innovation has been historical. That research began with the knowledge that such effects had occurred, and temporal and causal links with some past events were established. By and large the intention of such research has been to establish the complexity of social processes and the simple fact that such second-order consequences do take place. Such research has unquestionably been of value in teaching us respect for the ramifications of deliberate social action.

We have repeatedly expressed the point of view that previous research has probably overly dramatized the role of historical events such as exploration of the New World, the impact of the railroads, and so on. This thesis grew especially out of the work of Mazlish and his associates in their study of the railroads as an analogue to space exploration.[2] Our argument has not been that second-order consequences did not take place, but rather that their causation was more complex and more moot than generally assumed, that key events were only part of ongoing social trends.

However, this reasoning can be turned about, and we could just as well argue that the complexity of consequences has been as much underestimated as has been the complexity of causation. Rosenbloom's work on technology transfer offers good examples of this complexity of consequences. Aerospace technology can affect civilian technology in a myriad of ways. Some of these ways are exceedingly difficult to trace. Analogical thinking may be stimulated in a technologist in a consumer goods industry, who in his turn will produce an innovation that bears little surface resemblance to the original stimulus to his thinking. Or an aerospace technologist

---

[2] Bruce Mazlish, ed., *The Railroad and the Space Program: An Exploration in Historical Analogy* (Cambridge, Mass.: The M.I.T. Press, 1965).

may carry generalized technological competence with him when he transfers jobs to a civilian industry.

If we are indeed interested in understanding the full range of impact of aerospace technology on the civilian economy, we cannot do so by trying to trace this impact outward from its point of origin. What we would probably have to do is draw a sample of innovations introduced in a given time period, and trace their origins backward. To my knowledge no existing historical study has been based on a systematic sample of events that *might* have been caused by some large technological innovation to find out which in fact were so caused. This is the basis for my contention that we have underestimated the range of such consequences.

Clearly, at some point concern over our understanding or anticipating the full range of second-order consequences of either the space program or any other event becomes academic — and probably not even interestingly so. The potentiality of such complexity is limited only by the power of our imagination to conjure up new possible relationships, and additional waves of more remote impact. Pursuing the issue beyond a reasonable point can only create an image of the world that would paralyze action. It is at this juncture that one is better advised to become empirical, to act and depend upon an adequate system of feedback to report those consequences that are of sufficient importance to warrant an adjustment of course.

The Locus of Responsibility

The most fruitful way to examine an assumption is to act on it, or to act on it vicariously by imagining where it would lead one. We have done this with respect to the notion that an agency such as NASA has responsibility for the second-order consequences of its pursuance of its primary mission. The result is that which a reasonable man would have anticipated, namely, that for various reasons this responsibility must be limited.

An agency such as NASA has a primary mission assigned to it by the larger society, and its first responsibility is obviously to that primary mission. If it requires manpower or other resources that are in short supply, it has not only the right but the mandate to pursue such resources in an adversary fashion. If the impact is indeed adverse, this is a signal to the Congress and/or the executive branch that the missions it has assigned to the totality of agencies are in conflict with each other. Either the missions must be readjusted, or some additional effort must be launched to handle the consequences of the conflict — e.g., a program in manpower training may be inaugurated.

In practice an agency cannot always limit its concerns to its primary mission. At a minimum, an energetic pursuit of the primary mission without concern for adverse second-order effects can create an image of callousness and evoke an intolerable flow of criticism. In order to preserve its right to continue the pursuit of its primary mission an agency such as NASA must obviously maintain a reasonably favorable relationship to its surrounding environment. Even if other agencies existed to handle all second-order consequences of NASA's actions, there are situations in which it is symbolically important that the primary agency should manifest its concern. Certainly a clear-cut case would be the assumption of some responsibility for the impact of NASA installations on local communities. Or, on the positive side, it has been symbolically as well as practically important that up to this point NASA showed concern with the conversion of aerospace technology to civilian use. Whether this could continue to be so is another matter. It is clear that to the present there has been a strong expectation that NASA would accept this responsibility.

However, with the widespread growing concern over management of second-order consequences, there is and will continue to be an increase in the assumption of *general* governmental responsibility for this role. We find this already manifested in the development of local, state, and federal instrumentalities to deal with such phenomena as water, air, and noise pollution. Rosenbloom makes a strong case that management of technology transfer be the responsibility of some agency other than NASA both because of the magnitude of effort that a proper transfer program would involve, and because such a function would be equally appropriate to non-NASA technology.

At the time of this writing there is circulating in Washington the concept of a "Fourth Branch" of the government which would make independent evaluations of governmental programs and of the state of the society. It would be within the scope of this concept that such a "Fourth Branch" would recommend appropriate instrumentalities for the handling of many sorts of second-order consequences flowing from more than one program and/or lying outside the capability and reasonable responsibility of the program that produces them.

This matter is one on which some public discussion would be fruitful, and the discussion should include a consideration of the extent to which the public and its representatives believe specific programs should be responsible for objectives beyond their primary mission. The discussion should also consider the possibility of some of these functions being filled by nongovernmental institutions. There is precedent for communities and for private individuals having contracts with private firms for the removal of garbage, trash, and so on. There is no reason in principle that private firms could not handle problems of pollution or of social adjustment in rapidly expanding communities. NASA has already contracted with nongovernmental agencies for aid in technology transfer. Business firms are training Job Corps candidates; there are bills before Congress to draw profit-making institutions into solving problems of the ghetto, and so on.

While the policies associated with handling second-order consequences must be set by public agencies, there is no reason to believe that the execution of those policies should necessarily involve an endless proliferation of governmental activities and agencies.

The Space Program and Society: An Illustration

In suggesting that the management of second-order consequences be a matter of limited responsibility for an action agency such as NASA, and because NASA had a very loose mandate to deal with second-order consequences in the Act establishing it, I also indicated the larger societal responsibility for handling such affairs. Here I would like to take a single example, one to which Chapter 8 was devoted, that of scientific and technical manpower, to illustrate the range of broader societal problems that were illuminated under the stimulus of exploring the consequences of the space program.

A consideration of the problem of scientific and technical manpower in the Space Age proved to be a window through which we could see some possibly profound trends in society, which are associated with the increased demand for education in the occupational sphere. We were not the first to note the increased difficulties that this was beginning to pose for the young men (more so than for the young women) of underprivileged groups. Even when aspiring to jobs on lower levels than those of scientist and engineer, or even technician, they were beginning to suffer from lack of education and associated abilities. The picture of the evolving society is that of one even more demanding than that of the present, and unless the various elements of the poverty program are successful, this group may drop out of school at an accelerated rate.

Somewhat more novel was the picture that emerged at the boundary between the technician's job and that of the scientist and the engineer. Most of the elements of this picture were not new, but they came together to give a sharper image of a potential cleavage point in the occupational structure. Perhaps the technician's role is uniquely designed to illustrate the nature of this cleavage point, since it is explicitly defined as *not* professional, yet it demands most of the skills and values of the professional. Moreover, since it places the job incumbent in daily contact with the person he aspires to be, all these strains are accentuated. This sharply drawn picture, while it may almost be a caricature, could alert us to similar situations arising at other points in the structure where formal requirements differentiate adjacent jobs, whether vertically or horizontally. It is frustrating not to be able to move up, but it may be almost as frustrating not to be able to take a job for which one is equipped at the same level. The urgency of this consideration lies not only in satisfying the individuals involved, but in the necessity of providing an adequate quality of manpower in a rapidly changing work situation.

When we first approached the problem of manpower planning, we were impressed with the difficulty of predicting both the demand and the supply of people for various jobs. In the end we were convinced that precision of prediction is not as important as it had seemed initially. This becomes clear if we consider the problem of educating the new generation. Increasingly it becomes less relevant to anticipate the nature of the future work force and attempt to train people for specific entering jobs, since it is clear that to train for entering jobs is generally to cripple the person's chances for subsequent advancement. More and more he needs some broader range of skills, and particularly the skill to learn. It makes more sense to consider the level of education we can achieve for the population as a whole and train it to that level.

This is not to say that all persons have or aspire to the same level of competence. But we ought to make the attained level as high as possible both so that we may afford our people that which is due them, and so that we have a highly competent work force. If the economy continues to grow at its recent rate, and if the required skill levels continue to escalate, then we may abandon our traditional concern of having an oversupply of trained persons. There is more reason to worry about an oversupply of untrained, unskilled people.

Furthermore, we are now faced with the need for continuing training and education. Old dogs can learn new tricks and in our evolving economy they must learn new tricks. Regrettably our systematic studies of how people learn have generally been confined to animals, infants, and adolescents. We need to understand how people in their middle years learn, and then help them to do so.

Some Final Thoughts

In the preceding passage, I outlined how an attempt to understand one of the problems associated with the space program forced on us a reconsideration of many of the problems of the society which, it was obvious, could not be handled by the space program. In so doing I repeated or implied themes that have run throughout this book. The broadest of these themes is that NASA and its problems are so intricately tied to the broader trends in the society that the isolation of "NASA impact" in the over-all is virtually impossible, and/or to look at the impact of NASA as something entirely distinct from the broad impact of modern, heavily technical, innovative industry is to miss the point.

However, it would be equally serious to miss the point that NASA's visibility, and the symbolic role of the space program in the modern technological effort, has played an important role in stimulating inquiry into problems that concern the society as a whole. Furthermore, the space program in identifiable ways is shaping our ways of thinking and of living. Already space science has enlarged our knowledge of our own earth and of our universe. Space technology has influenced weather forecasting and television communications. The notion that men can travel in space — so drastic an idea only a decade ago — is now a commonplace.

While I would defend our decision to "research the researchable" as a way of getting started, it seems obvious that many of the side effects of the space program fall outside our very limited net, and some of them would be most difficult to measure. For example, the increase in technological competences flowing from the space effort has an impact on the economy in many ways so diffuse that one could never estimate with any precision what the contribution, both positive and negative, has been.

There will, in these and all other things, always be limits on what we can research systematically, or on what we find it worthwhile to research. Beyond these limits will be matters on which judgment and speculation must be invoked, because they are matters about which we care even if they are only imperfectly understood. Our intention in making a few forays into more systematic inquiry was not to deprecate those necessary intuitive attempts at trying to understand what the Space Age will mean to us but to better define where speculation might stop and data gathering begin.

Appendix

Space Efforts and Society

A Statement of Mission and Work

A Document of
The Committee on Space Efforts and Society
of the American Academy of Arts and Sciences

Boston, January 1963

Preface

In April 1962, the National Aeronautics and Space Administration (NASA) gave the American Academy of Arts and Sciences a grant of $181,000 for the purpose of studying, over a period of approximately two years, the relationships of society to efforts of massive technological innovation, with special reference to such efforts as the civilian space program. NASA phrased this grant in broad terms with the expectation that the committee administering it would, as its first task, develop a clearer conception of what was to be done, and would support its proposals with a well-articulated rationale. The document which follows is an attempt by this Committee to develop such a statement of mission and supporting rationale, and to propose activities which flow from this statement.

Our report is divided into three parts. In Part One, "The Mission," we present our view that the most urgent and worthwhile issue for study under the grant is the problem of living with technical change, especially dealing with the second-order consequences which will inevitably result from such change. We further note that the goal for study in this area should be to develop devices for anticipating, detecting, evaluating, and acting on such second-order consequences. Finally, we conclude that the limitations of time and money, plus the logical sequence of the work to be done, combine to narrow the mission of our Committee to activities relevant to anticipating and detecting second-order consequences.

In Part Two, "Developing a Strategy," we discuss two important questions that must be evaluated before specific research can be proposed. First, what is the current state of the art in the theory and methods of anticipation and feedback, and of what value are these current techniques and concepts to the Committee's concerns? Second, what activities should be preferred over others? By giving our answer to these questions, we provide a framework for the activities presented in the next section.

In Part Three, "The Work Proposal," we present the specific activities that we have chosen to fulfill our statement of mission and its rationale. The first section, "The Nature of the Task," presents three problems which we believe require general investigation. The second and third sections, "Devices of Anticipation" and "Detection of Side Effects," present specific suggestions for research on the impact of the space program.

The breadth of NASA's mandate to the American Academy would allow this Committee to propose any number of different programs. In fact, it is the very range of things that could be proposed under the grant that made it imperative for us to establish a mission and to set standards for choosing among the activities to be undertaken in fulfilling that mission. Consequently, a reader will find that our particular approach may have omitted a specific piece of research or writing that appears highly desirable in its own right — but which we determined did not fit our mission or rationale. We hope, however, that the over-all coherence of our proposals will be sufficiently attractive to offset the loss of any individual activity.

Though the *focus* of our attention is the American civilian space program, stating this focus does not define the boundaries of our activities. Both the nature of human behavior and research activities will cause us to do some work in areas that are not clearly "civilian space activities." For example, it is difficult to distinguish the impact of the NASA program from that of the Russians or the American military; it may be easier to do research in a different area of human behavior analogous to the civilian space program, than in an area of direct present concern. Similarly, research or study activities may have to pursue ideas across the "civilian space research" boundary into problems of greater generality. For example, studying the problems of transferring "space" technology to other areas of endeavor may require a broader investigation of technological transfer from one enterprise to another. We are, of course, fully aware that the generous terms of our grant have already anticipated the possibility that our inquiries would need to be broad. We pay tribute to NASA's farsightedness in sponsoring our study in this way.

At the same time we must remember that our inquiry, as an activity of the American Academy of Arts and Sciences, is not a mere technical service to a given agency. Rather, this is an inquiry into a general problem of which the civilian space program is an important example, and a general problem in which NASA has interest. The Academy undertakes this inquiry under the assumption that the significance of such study extends well beyond the civilian space program, and in the belief that the space program must be evaluated in terms of national and international considerations.

Finally, this document was prepared by the Committee Chairman with the assistance of critical comment from members of the Committee and the Executive Director; and benefited from the valuable stimulus of documents on the topic prepared during the summer of 1962 by Bertram Gross, John Seeley, Sumner Myers, Edward Furash, and Lewis Dexter.

THE COMMITTEE ON SPACE EFFORTS AND SOCIETY

Raymond A. Bauer, Chairman
Francis M. Bator
Saville R. Davis
Donald G. Marquis
Don K. Price
Walter A. Rosenblith
Earl P. Stevenson
Arthur E. Sutherland
Lewis A. Dexter, Executive Director

Boston
January 1963

Highlights and Table of Contents

B. *Dealing with Side Effects:* Even though man has been conducting his affairs with reasonable effectiveness without good devices for anticipation and feedback, such *ad hoc* behavior is not the optimal way of proceeding; and the greater impact and speed of modern technological change demand anticipatory planning lest there be more victims than beneficiaries

A. *The Past, Present, and Future of Such Work:* Our review of the history and work to date on the social aspects of space exploration poses the question of how to move from a situation in which "interesting ideas" are posed to one in which a reasonable amount of scholarly work is actually taking place

B. *Some Criteria of Exclusion and Inclusion:* The criteria we used in developing the statement of mission and in selecting the particular work to implement that mission: (1) a desire to support unique and distinctive contributions; (2) a desire to support strategic contributions building upon the existing state of the art; and (3) a desire to maintain coherence of effort

C. *Development and Research:* In the light of the criteria posed above, much of the research that needs to be done might more appropriately be labeled *development* rather than research. The objectives of such development work are: (1) establishing the existence, importance and nature of the problem; (2) defining problems more adequately; and (3) producing substantive contributions

D. *Some First Steps:* Three questions we will use when determining whether to undertake or support an activity

Part Three: The Work Proposal

The Work Proposal discusses a list of projects we believe will achieve the objectives presented in our statement of mission. The full scope of this work proposal exceeds what is practical under the present grant. It is, therefore, a list from which we will choose to do some things with a degree of thoroughness, to commission essays on others, and to encourage work on still others. The first section, "The Nature of the Task," presents three problems which we believe require general investigation. The second and third sections, "Devices of Anticipation" and "Detection of Side Effects," present specific suggestions for research on the impact of the space program

A. *The Problem of Social Indicators*
B. *The Space Program and National Goals*
C. *Feedback into Organizations*

A. *Historical Studies*
  1. The social impact of technological change
  2. Events which changed national self concepts or man's image of himself

Part One: The Mission

I. The Problem: Living with Technical Change

A. Second-Order Consequences

In the conduct of human affairs, our actions inevitably have second-order consequences. These consequences are, in many instances, more important than our original action. For example, the automobile is said to have changed our leisure life and sexual mores; generated suburban living and changed our patterns of home ownership and retail distribution; modified the use of roadbuilding as patronage and created new special interest groups.

These second-order consequences present problems of: (1) anticipation and/or early detection; (2) evaluation; and (3) action. It is the concern of NASA with such second-order consequences of the space program that prompted its grant to the American Academy of Arts and Sciences. Such a concern is stimulated by the hope that the activities of NASA can be carried out in such a fashion that the net total of secondary effects can be optimized — within reasonable limits. Furthermore, anticipation and/or early detection of second-order consequences could probably affect the choice of space goals or the manner in which they are pursued.

B. The Isolation of Technical Problems[1]

Technical changes have proved to be particularly explosive sources of social, economic, and political change. (Henceforth, we will use the one word "social" to embrace all these forms of change.) This explosive potential is inherent in "technical" change. Nevertheless, a problem posed as a "technical" one tends to beg

---

[1] In this context, the word "technical" is used in the broad and relative sense and is meant to include "technological." Thus, a legal problem would be a "technical" problem when it was assumed that a clear-cut, unequivocal answer could be found if only the investigator were bright enough and understood the law well enough. When "technological" as a narrower concept is intended, it will be so noted.

199

the question of consequences because of the very way in which it is customarily phrased. People start, generally, with the assumption that there is an *agreed upon* need, and the task is simply "technical," e.g., one of devising means for meeting that need. Given such a conception of the problem, there is a bias toward evaluating possible actions in terms of whether they will meet the "agreed upon" need, and a bias against thinking in terms of the range of effects which will flow from such actions. Furthermore, while there is usually consensus that there is "a problem," agreement on the exact nature of the problem is often a difficult task! Therefore, when an apparent solution to the apparent problem has been found, few of the parties involved are much disposed to reopening the issue unless the need is clear and urgent.

These biases are not absolute, but they are re-enforced by other factors. One is that the technical solution to a specific problem tends to seem unequivocally demonstrable. Within reasonable limits one can usually say quite comfortably that a technical problem has been solved. However, the second-order consequences are inherently more difficult to anticipate, and to the extent that they are anticipatable they are difficult to evaluate. Particularly when the "technical" problems are specifically technological, our yardsticks are much more clean-cut than those used in assessing the value of social change. For example, at this very moment in our history there is little agreement as to what will be the impact of automation on over-all employment, skills required, the national economy, or the world economy. Even if these outcomes could be predicted with precision, we would not have an agreed upon yardstick by which to judge them. On what metric do we compare a given level of unemployment among unskilled, young, Negro men with a given rate of increase in productivity? How do we project such a comparison into the future? How far into the future and on what assumptions?

Given these circumstances, and a civilization which places a high value on science and technology, and in which progress tends to be identified with technological change, it is no wonder that concern for the consequences of technological change is frequently met with some such rejoinder as, "But you can't stand in the way of progress!" To this is added the observation that innovation has always been greeted with doubts, but that Western civilization has continued to make material progress. Such rejoinders are often a dismissal rather than a solution of the problem. But at least it can be said in their defense that they help avoid paralysis.

C. *The Difficulty of the Task*

The above paragraphs outline a few of the reasons why men are often reluctant to consider frontally the second-order consequences of technological innovation. It is by no means clear, however, that even the most forthright facing of the issues can bring about an appreciable improvement in our affairs. The history of the social impact of many innovations (e.g., the automobile) suggests a chain of events which proceeded inevitably, irresistably, and irreversibly. Given this picture of inexorable consequences, one may be particularly struck by the faith generated by the space program: that on some sides there is a simple, unquestioning idolatry of the most rapid possible pace of technological innovation. While there are many grounds for regarding such innovation as desirable, it can scarcely make more simple the task of guiding our activities toward the service of all our common values. There is a strong probability that massive technological innovation of the sort involved in the exploration of space is a virtual guarantee of rapid social, economic, and political change of an unpredictable and uncontrollable nature. The appropriate response to such a situation might be a modified form of social Darwinism. Technology may be viewed as producing a large volume of essentially random social mutations, and our future as a race may not lie in deliberate planning, but in developing the most adequate possible mechanisms for detecting and selecting among such mutations.

Some discussions of this problem by technologists see it condescendingly, as a result of the slow rate of social innovation in our present era, as though other eras have done conspicuously better on this score. This indeed puts the cart before the horse. If appropriate measuring devices could be agreed on, we might well conclude that the present rate of social innovation is exceedingly rapid, but also that the problems generated by technological change are unprecedented by several orders of magnitude.

All of the foregoing is not a preachment of despair. It is on one hand a statement of the importance and the difficulty of the problem to which NASA has turned its attention. On the other hand it is an affirmation of the fundamental and relatively untouched nature of the over-all problem with which we are faced.

II. Anticipation, Detection, Evaluation, Action

It was said above that the appropriate incorporation of second-order consequences into the conduct of affairs involves the following problems: (1) anticipation and (to the extent that they cannot be anticipated) early detection; (2) evaluation; and (3) action. For purpose of exposition we discuss them in reverse order.

A. *Action*

By "action," we mean doing something about the problems or opportunities which present themselves. This is not to imply that something can in fact always be done. It is rather a prosaic observation to the effect that if we know about the existence of a problem or opportunity we are likely at least to explore the possibility of doing something about it. We are giving secondary priority to corrective action in the plans for the work of the Academy Committee and concentrating our attention primarily on problems of anticipation, detection, and in a specifically defined sense, evaluation.

There are three reasons for giving second priority to problems of action:

(1) Corrective action must of necessity follow the detection and evaluation of the effects of our actions. Study of these latter problems is by itself a sufficient task.

(2) More importantly, it is not clear in anticipation of the event what is the proper agent for action. Some opportunities and difficulties created by the space program may properly fall in the domain of agencies such as local school boards rather than of NASA or any Federal agency — or might demand the creation of new agencies or techniques. Understanding of this dimension of the over-all problem is an essential part of our task.

(3) *Responsible* action proposals must be concrete, taking into consideration the *complexities* of the specific circumstances of action. This is saying something beyond (1) above, that action follows detection and evaluation. It is saying that once detection and evaluation have been accomplished, responsible proposals for action would involve resources beyond those available to this enterprise. We will not, however, hesitate to outline what seems to be the set of possible actions without choosing among them, whenever our analysis is sufficient to warrant such suggestions.

B. *Evaluation*

Manifestly the relationship of the NASA program to our national goals and to other human values is a problem of central concern to all groups in our society. It is precisely because evaluation is so vitally important that we would like to define our exact relationship to this problem. We are not lacking today in either quantity or variety of evaluations of the space program, or of any of our national activities. The major contribution we can make is not to contribute another wise pronouncement, but to see if we can make a contribution to the general level of evaluation. It is neither our mission to legislate values, nor to bootleg in unannounced our own version of such values. Our mission is rather to pose the questions of evaluation as sharply as possible so as to encourage and facilitate the making of decisions of evaluation. Yet, as we are continually reminded, the values of the nation are ultimately the values of the people themselves, and they must make these decisions. The transfer of these values to guides for action is a significant problem in itself.

Our relationship to the problem of evaluation will in no sense be passive or evasive. One of our proposals is for a study of the ways in which the interrelationship of national goals, including space exploration, may be worked out. In our judgment, virtually all studies of national goals have assumed that there is little difficulty for reasonable men in working out the conflicting demands of various goals, or have indicated a priority of goals that reflected the preferences of the group making the judgment. Even a priority list of goals is not in itself an adequate guide for the complex trade-offs that must be made in specific instances. Hence, it is because we take this problem so very seriously that we will attempt to discipline ourselves to make a distinctive contribution, by making a limited one.

C. Anticipation and Detection

With all the qualifications and for all the reasons entered above, *we will direct our efforts primarily at the technical problems of anticipating and detecting the second-order consequences of massive technological innovation, particularly as exemplified in the civilian space program.*

The purpose of detection is to provide "feedback" to some action agency. We introduce the term feedback here because of the general notions of planning, action, and control with which the term is associated. Recognizing that its use in the social sciences is generally somewhat different than its more precise meaning in electrical engineering where it was developed, we will define this term for our purposes as, "the detection and reporting of the consequences of one's actions with sufficient rapidity and in such a form as to be a guide to corrective action."[2] The addition of the notions of time and that the form of detection and reporting shall be appropriate to conceivable remedial actions is an important addition to the mere idea of detection. What we are concerned with, then, is not a merely academic enumeration of lamentable catastrophies and missed opportunities, but the detection of things about which something might be done, and which might be reported soon enough and in such a form that something might be done about them.

This, however, is looking far down the road toward the day when feedback mechanisms might be built systematically into the nation's space efforts. For practical purposes, we will be occupied with *detection*, while keeping these other considerations in mind.

The companion term "anticipation" is also broader than another word which might have been used, namely "prediction." The distinction we would make is that between a reasonably precise knowledge of the probability of a given event happening, and a sufficient awareness that something of a general sort might happen, that if it does happen we are not caught by surprise. In planning it is often possible to make provision for handling some consequence even though it can be foreseen only in very general terms.

There is another sense in which the distinction between "anticipation" and "prediction" is vital. Prediction is an enterprise which places a primary emphasis on the probability of a given event occurring in the future. Prediction *per se* should not be concerned with the *importance* of various possible future events — though as a matter of fact the probability and the importance of the events often get confused in people's minds.[3]

If we succeed in keeping the issues straight, we may "anticipate" the possibility of some future event of great importance, though other outcomes might be more probable. Yet, its importance relative to other possible outcomes might be so great that we would be wise to act as if it were going to occur. Thus, when we embark on an automobile trip, the probability of encountering an accident is low, but this is no reason for not using a seat-belt. Or, viewing the matter in reverse, using a seat-belt does not constitute a prediction that one will have an accident.

---

[2] This definition is adequate to our purpose here. However, as shall be suggested in one of our work proposals, a further specification of the concept of feedback as developed in electrical engineering with a listing of the analogies and disanalogies in human organizations may be worthwhile.

[3] For a discussion of this issue see Raymond A. Bauer, "Accuracy of Perception in International Relations," *Teachers College Record, 64* (January 1963), 291–299.

This distinction takes on particular relevance in the light of the proclivity of social scientists and social critics to write "cautionary tales" such as *1984* or *Brave New World* as though they were the most probable estimates of the way things will be. They are often well warranted as "forecasts" of *possible* contingencies. But they should not be confused with "predictions" of what is most likely to be. It is an empirical question as to which question is more important, but in planning for the future the important but improbable outcome is often more relevant than the unimportant but probable one. "Anticipation" is a word that directs our attention to the full range of possible outcomes, while "prediction" directs our attention to the most probable.

Part Two: Developing a Strategy

I. Anticipation and Feedback[1]—The State of the Art

The end state toward which we are oriented is that an enterprise of the size and importance of NASA should have incorporated in it a mechanism of anticipation and feedback that would enable it to guide its actions with respect to its second-order social effects.[2] It is our conviction that such an end state will not be achieved in any single decisive step, but will rather be reached by a sequence of approximations, the first of which will be very crude indeed — judged against the complexity of the task as we can even now comprehend it.

A. *Indications of Change*

Considering the confidence with which social critics of this and past eras have commented on trends in society, one might assume that the basic task of being able to measure "good" and "bad" trends in society is well in hand. In point of fact, this problem has scarcely been approached. The field of economics is an exception against which other areas may be evaluated.[3] Economists have, over a number of decades, developed a series of indicators which are used to evaluate and anticipate trends in the development of the economy. Economists differ in their evaluation of the meaning and implication of such indicators. However, relatively speaking, there is consensus on *what* should be observed and how it should be observed. *Some* isolated trend series have been developed in vari-

---

[1] As indicated above, feedback includes both the detection *and* the reporting of effects in the appropriate form to the appropriate agency.

[2] Such "actions," it may be stated, would include reporting of opportunities and problems to other agencies when NASA was not the appropriate actor. Such actions, furthermore, could extend to modification of NASA's *primary* goals in instances where second-order consequences were unduly detrimental.

[3] It is an interesting point that the United Nations document, "Report on World Social Situation," U.N. Department of Economic and Social Scientific Affairs, Sales #61.IV.4, 1961, deals chiefly with *economic* indicators.

ous other areas of society, notably in fields such as public health. To discuss the technical difficulties of most of these series, e.g., the incomparability of data from various time periods, might seem captious, since what is initially important is the lack of consensus as to what should be measured. This lack arises in part from a deeper lack of consensus — lack of consensus on goals, purposes, and the nature of society.

Probably some reasonable agreement could be reached on some range of goals and values on which there is consensus. Let us at least assume this. A new series of problems develops immediately. Many goals and values are of such a level of abstraction that they give us little guidance as to what to observe. Take as examples "national unity," or "the development of the fullest potential of the American people." These are values with which few Americans would disagree. Yet there are wide variety of events which one might observe and which bear on these values, and they would not all be moving in the same direction at the same time. Consequently, the resultant observations would be hard to coordinate.

Even the strictly technical problems are formidable. An economic index such as cost of living must be constantly readjusted to account for the changing composition of the standard market basket. In a similar manner, even so simple a phenomenon as change of educational level of the population cannot be evaluated properly without taking into account differences in the content of the curriculum, the uses to which education is put, and the meaning of extending education to successively less intelligent segments of the population. The difficulties of developing indices of social pathologies for, e.g., crime, mental illness, mental deficiency, divorce, and suicide are enormous because of the shifting definitions of such categories.

It might be assumed that when there is agreement on our goals, and values are agreed upon, we can decide what observable events will be "surrogates" for these goals and values,[4] and we have solved the technical problems of measuring and comparing them over time, that our job is done. But, this is not so. If we have solved all the foregoing problems we will then be in a position, hopefully, to evaluate the trend of events *up to the present time*. But policies are set for the future! Put most simply, if an observation is made today and assessed as to what it means tomorrow, this assessment must

---

[4] Among the few scholars who have tried to develop schemata for such purposes, we may mention Charles E. Merriam, *Systematic Politics*, 1945, and his student, H. D. Lasswell, *World Politics and Personal Insecurity*, 1935.

be made on the basis of the flow of events between today and tomorrow — on some assumption as to what tomorrow and the intervening time will be like! Thus, an increase in desire for technical education among the young may be regarded as a "good thing," but we can be confident of its desirability only to the extent that we are confident that the need for such skills will be important at the appropriate time in the future.

Viewed in this way, social indicators are simultaneously devices for prediction, and also yardsticks of progress which can be used only on the basis of assumptions about the future. (It seems to us impossible to conceive of any concrete situation, in terms of the total nature of society at present no matter how desirable, that could not prove undesirable under some conceivable future development of society.)

In outlining the complexity and difficulty of this topic, we are neither preaching despair nor promising an "answer." What is at stake is the whole difficult task of how man goes about guiding his own destiny. Surely this goal will be achieved slowly and by approximations. Even a "first step" so modest as defining the scope and nature of the problem would be progress.

B. *Dealing with Side Effects*

It might be appropriate at this point for the reader to object impatiently that we are posing impossible conditions; that man has been conducting his affairs with reasonable effectiveness for some time, and without such consensus and such knowledge. But there have been certain features of our large-scale enterprises in the past that represent exactly what NASA is trying to avoid. There is drawn up a plan for urban renewal, or a plan for development of a socialist economy such as that of the U.S.S.R.,[5] and progress is evaluated almost exclusively in terms of its primary objectives. The most obvious of second-order consequences (e.g., displacement of population) are provided for — often it must be said to the extent that the people affected are able by their complaints to prejudice the primary objective. Feedback is rudimentary. Desirable second-order consequences are windfalls, and seldom the result of deliberate action. Undesirable second-order consequences are dealt

[5] The pattern sketched here was especially characteristic of the planning and execution of Soviet economic development. "Second-order consequences" were generally ignored until they grew to major proportions, at which point they would be made a major objective to be pursued, in turn, without much regard for other consequences.

with on an *ad hoc* basis, and usually only when they have reached or obviously threaten to reach, fairly major proportions. In other instances, it is the "victims" of these consequences, rather than the initiators, who are left to cope with them (the initiators may be, but rarely are, the immediate "victims"). The fact that we have survived and made certain progress to date indicates either that this is a viable, though less than optimal way of proceeding, *or* that we are the lucky remnants of a rather sloppy selection process.

But, this is precisely the situation from which NASA is trying to extricate itself. Exploration of space is unique as an enterprise in which for many years the second-order consequences (e.g., the impact of the program here on earth) will probably have much more obvious effect on man's affairs than will the success of the primary mission. Departure from traditional ways of handling side effects of one's action is a precise parallel to the systems approach that dominates modern technology of which the NASA program is a prime example.

It is probably already clear that a particular philosophy of planning is implied in the foregoing pages. An enterprise as complex as NASA cannot operate on the basis of a *rigid* projection of events into the future, but on the other hand, because of its extraordinarily long time perspective it must operate with *some* form of projection of events into the future. To avoid becoming the victim of its own commitments it must incorporate into its plans provision for a wide range of contingencies combined with an active sensing mechanism for effecting feedback. Our intention in making these statements is not to usurp any planning function of NASA but rather to state the bases on which we will direct our own concern with the anticipation and detection of the social consequences of NASA's actions.

II. Some Orienting Considerations

A. *The Past, Present, and Future of Such Work*

Interest in the social context and consequences of space exploration is, of course, by no means a new phenomenon. The history of some relevant streams of thinking is quite long. Concern for social aspects of space exploration in science fiction dates back at least several decades. Work of a more empirical nature, such as public opinion studies of reactions to "flying saucers," extends back at least a decade. The inauguration of the formal era of space exploration by the launching of the Soviet Sputnik in 1957, produced some increase of interest, and a certain amount of research (mainly in the form of public opinion studies). Such interests and activities have continued to the present time.

It is important not only to note that concern with the social aspects of space exploration has a reasonably long history, but also to note that such interest has failed to develop really effective momentum in the academic community. Isolated individuals and groups continue to write and work on space related social problems. However, a high proportion of this work is done on an avocational basis. It is interesting for example, that there is a "Space Committee" of the Society for the Psychological Study of Social Issues, an active, self-organized, informal group of psychologists and sociologists living almost exclusively in New York City, mainly employed full time in commercial market research or similar organizations.[6]

The most substantial contribution which has been made was the work of a NASA supported study group, organized by The Brookings Institution and directed by Donald N. Michael.[7] This group met over a period of two years, and received a series of staff papers on the social implications of various aspects of the space program as it was then envisioned. The result was a 270-page document, densely packed with information and speculation on the implications of the space program, together with thousands of suggestions for research. Whatever over-all evaluation one might make of the Brookings report, it was certainly not lacking in coverage. Since its submission to NASA in November 1961, relatively few ideas have been introduced into the discussion which are not anticipated in some form in the Brookings report.

This latter circumstance is of extreme importance in considering the present stage of activity. In the summer of 1962, another study group convened at Iowa State University under the sponsorship of the National Academy of Sciences/National Research Council.[8] The report of the Iowa 1962 group is more sharply focused than is the Brookings report. The mandate of the Brookings group led it to

[6] One of the few highly visible examples of writing on social aspects of the space program is an issue of the *Journal of Social Issues* organized by and largely written by this group: cf., "Social Psychological Implications of Man in Space," *Journal of Social Issues, 17,* No. 2 (1961).

[7] For the report of this work, see Donald N. Michael, *Proposed Studies on the Implications of Peaceful Space Activities for Human Affairs,* U.S. Government Printing Office, 1961, available from The Brookings Institution and from Universal Microfilm Service, Inc., Ann Arbor, Mich., OP 17840.

[8] See "Some Social Implications of the Space Program" in *A Review of Space Research,* National Academy of Sciences/National Research Council, Washington, D. C., 1962.

emphasize coverage. Accordingly, it contained so many suggestions of what to do, that as a practical matter it was less likely to stimulate research than would a more limited list of suggestions. The 1962 study group makes a more limited number of suggestions for research and action. That circumstance may make it more probable that some of the suggestions will in fact be implemented.

What is of major concern, however, is that the 1962 Iowa State group was grappling with the problem at about the same level as that of the Brookings Institution group. As is appropriate for a study group, the emphasis is largely on work that other people or institutions ought to do, or could do. This too, was the major focus of the Brookings group. There is very little evidence either in the 1962 report, or in the literature in general that following the Brookings report any considerable group of scholars actually did any serious work on space related problems with social implications.[9]

The level of speculation on the social implications of space exploration has been generally provocative, stimulating, imaginative, and usually well informed. This is true of the Brookings group, the 1962 Iowa conference, or the suggestions which the present Committee received from the Fellows of the American Academy of Arts and Sciences.[10] The major question, therefore, is one of how to move from a state of "interesting ideas" to one in which a reasonable amount of scholarly work is actually taking place to test and capitalize on some of these ideas. This is an over-all strategic consideration in our thinking. It is not sufficient that our own work shall in itself be worthwhile, but that it be selected, conducted, and presented in such a way as to increase the probability of further work on space related social problems.

One essential requirement for implementing this strategy is that the work proposed be phrased as closely as possible to traditional problems of current interest to relevant scholars. It seems as though

[9] Exceptions would be work on civilian applications of military and space technology (Project Carmrand of the National Planning Association), and work on manpower problems sponsored by the National Science Foundation; we hope to issue in the spring of 1963, a checklist of items that have come to our attention that directly refer to the space program. An excellent example is, "A Pilot Research Study to Determine the Patterns of Communication between NASA and Groups Within the Scientific and Professional Community," United Research, Inc. (Cambridge, Mass., February 1961).

[10] These suggestions from about 200 Fellows of the Academy have been summarized by Geno A. Ballotti in a document available from the Committee at 12 Garden Street, Cambridge, Mass. The suggestions came in the form of replies to a letter sent on the behalf of the 1962 Summer Study Group of the Space Committee of the Academy.

there may have been a major micalculation in some quarters in assuming that the inherent interest and importance of the space program would automatically attract first-rate scholars to work on space related social problems. (We have often heard the phrase: "But, it's the most important thing happening in the world today! Surely they are interested!") This assumption ignores the structure of the scholarly community with its established traditions of problems, career lines, and status. Mere topical interest or practical importance of a problem area is not likely to attract much worthwhile effort unless it can be related to the traditional pattern of problems.

B. *Some Criteria of Exclusion and Inclusion*

With this broad strategy in mind, we would like to record some of the criteria used in developing our statement of mission, and which will be used to select the particular pieces of work to implement that mission.

(1) A desire to support unique and distinctive contributions. Certain types of work would be done irrespective of our support. For example, there is a well-established and relatively well-founded tradition of macroeconomic studies on such topics as the probable impact of various patterns of investment on the over-all economy. It is virtually certain that such studies would be carried out independent of our own efforts. Consequently, *for our purposes,* such studies, important as they may be, are given low priority.

Similarly, other relevant studies were known to be already in progress. Thus, some are attempting to project the manpower requirements of the space program, and the effect of these requirements on competing manpower needs. Hence, our only proposal in this general area would be for methodological evaluation of the existing and ongoing work, which does not duplicate these efforts.

(2) A desire to support strategic contributions building upon the existing state of the art. Concern with the social implications of space exploration, as has been said above, is by no means virgin territory. Many different contributions to the discussion could be cited. All-in-all, much thoughtful informed speculation on possible and probable effects of space exploration has been diligently pursued. Our orientation is to take such previous work as a point of departure, and to support what we have called "next first steps" toward the ultimate objective — which may indeed be very distant — of optimizing the full range of consequences of the space pro-

gram. Our own proposals will be very modest precursors of this objective.

We react similarly to various pertinent areas of general knowledge, and propose to build upon the existing state of the art in each such area. Among these efforts are (1) examining current practices with respect to improving the selection and measurement of appropriate indicators of social progress; (2) researching techniques for selecting appropriate actions in a situation of multiple, conflicting organizational goals; and so forth. In such instances, we were moved by the belief that the state of the art has to be advanced if our general task were to move forward. This latter range of considerations is reflected in the section which follows immediately, "Development and Research."

A difficult but essential part of the job, is, we believe, to take much of the speculation that has taken place and turn it into doable research. Where other considerations were approximately equal, attention will be paid to the demonstration value of attempts to gather data on topics which might seem difficult to research. Probably the outstanding example of this is the research which is suggested to anticipate the possible effects of the discovery of extraterrestrial life.

(3) *A desire to maintain coherence of effort.* While the work statement which has been developed is marked by considerable diversity in content, this diversity was tempered by the intention that the various pieces of work should reinforce each other. As a result, many highly important individual topics were excluded. A good example of such an excluded topic is the potential role of space explorations in our relationship to the Communist bloc.

Readers of early drafts of this document have on occasion suggested that coherence of effort could better be achieved by concentration on one or two main topics. The surface plausibility of such urgings cannot be denied, and the policy we have proposed may in fact be proven ill-advised. Our decision is based rather on the strategy proclaimed above of building momentum in this area of inquiry rather than of accomplishing one major significant piece of work. The reasonableness of our choice should be supported by the section which follows immediately.

C. *Development and Research*

As our deliberations proceeded, it became clear that, in the light of the criteria stated above, much of the research that needed to be done might more appropriately be labeled *development* rather than research. Specifically, there are existing bodies of knowledge which have not been organized and brought to focus on the general problem to which we are addressing ourselves. For this very reason many of the specific work statements call for essays drawing on existing knowledge.

One general problem area may be cited as an illustration. To date, the contributions of research and development to economic growth and social well-being have concentrated generally on: (1) macroeconomic studies of the effects of investments of given orders of magnitude; and (2) technological evaluations of the potential value of actual or potential innovations. At the same time, the *actual process* of translating space and military technology into relevant civilian technology (as an example of the general and difficult process of introducing innovation into an ongoing organization) largely has been ignored. The introduction of innovations into organizations, in fact, has either been taken for granted, or been regarded with irritation when it does not take place. Here is a problem area in which there is considerable relevant knowledge that has not been drawn together or examined systematically.

The objectives of developmental work, of which the foregoing is simply one example, are:

(1) Establishing the existence, importance, and nature of the problem. This objective is in part prophylactic. A mere awareness of the importance and nature of the problem will often help avert mistakes. For example, if responsible persons are alerted to specific questions to ask about the introduction of innovation into an ongoing organization, their own common sense will suggest appropriate ways of facilitating the situation, or at least of not exacerbating it.

Over the long haul, it is perhaps even more important that the very knowledge of the importance of many of these problems will be an incentive for the commissioning of continuing research along such lines — and will help to focus attention on the formulation of key critical problems for extensive field work.

(2) Defining problems more adequately. Adequate definition of the problems will supply guidelines for further work. For example, there has been no systematic effort at a statement — to the best of our knowledge — of the difficulties of converting the full range of generally accepted national goals into operational policies and pro-

cedures. What observations should we actually make to decide whether or not the over-all welfare of our people is improving?

(3) Producing substantive contributions. Some of the work which is planned should produce knowledge of immediate practical importance. For example, a review of Japanese experience in adopting unfamiliar technological developments into local organizations might help in understanding the task of translating space technology to civilian use, as would a series of case studies of instances within our own country where such technological exchange takes place.

D. *Some First Steps*

Our task, therefore, is to devise a series of first steps in the direction of making it possible for such large-scale enterprises to better guide their actions with respect to the full range of their effects. The primary purpose of these first steps is not to produce substantive findings, but rather to understand better how one would ultimately go about the large task with which NASA is faced. This, again, is not to say that we will not produce substantive findings. As a matter of fact, an essential part of our mission is to demonstrate that useful substantive findings can be produced, and to use substantive findings to illustrate the nature of the problem of building a feedback mechanism into a massive technological enterprise. While it is expected that many of the concrete findings will be of immediate interest and use, any one activity ought to be justifiable primarily in terms of one of the three following questions: (1) Does it help us understand the nature of the task (e.g., a study of how to decide on what social events to observe in order to judge the social impact of the space program)? (2) Does it help us *anticipate* what the potential consequences of NASA actions may be (e.g., a survey of science fiction as a source for an inventory of plausible consequences, or an historical study of the impact of voyages of exploration)? (3) Does it help us learn to detect what the effects actually are (e.g., the tracing of the diffusion of new technology, or studies of space-impacted communities)?

Part Three: The Work Proposal

The Work Proposal that follows consists of a list of projects having promise of forwarding the objectives presented in our statement of mission. The full scope of this Work Proposal exceeds what is practical under the present grant. It is, therefore, a list from which we will choose to do some things with a degree of thoroughness, to commission essays on others, and to encourage work on still others. We hope that our projects will stimulate other scholars to pursue these matters further. The first section, "The Nature of the Task," presents three problems which we believe require general investigation. The second and third sections, "Devices of Anticipation" and "Detection of Side Effects," present specific suggestions for research on the impact of the space program.

I. The Nature of the Task

We are concerned with the anticipation, detection, evaluation, and reporting of social consequences of both the exploration of space and of the activities associated with it, such as the employment of large numbers of persons, the spending of large sums of money, the generation of knowledge and technology, etc. The inherent difficulty of the task would be sufficient by itself even if it were not compounded by the flood of social criticism which blithely glosses over vast areas of ignorance with confident statements about trends in "American values," "morality," "conformity," "welfare," etc. It takes no particular skills in prediction to anticipate that the space program will shortly be exposed to a flood of such criticism, probably both favorable and unfavorable, and probably, also, largely irrelevant. There would be a substantial prophylactic value in doing no more than making clear the difficulties of such evaluations at our present stage of knowledge, providing such a definition of the problems improved the public's capacity to evaluate this type of social criticism.

Our intention, however, is to accomplish more than a prophylac-

216

tic job. In this section of the work statement we shall propose general approaches to three aspects of the task: the problems of social indicators, national goals, and feedback into organizations. These problems, however, also run through many more parts of the work statement, playing a role in the consideration of specific aspects of the space program. It is expected that when the project is completed the more general and the more concrete approaches to the task will each illuminate the other.

A. *The Problem of Social Indicators*

We indicated in preceding pages that there is presently no consensus as to what phenomena in our society should be observed in judging whether trends of the society as a whole are favorable or unfavorable. Even in those instances where an operational criteria for individual variables (e.g., GNP as a measure of economic growth) can be agreed upon, there are usually difficult technical problems in making observations that are comparable for successive time periods. More typically, however, totally ambiguous criteria are interpreted to suit the convenience of the user; e.g., does a high suicide rate in Sweden reflect the weakness of a socialist society, or its ability to keep good statistics?

No short-term effort will produce a tidy resolution of this problem. However, the field is ripe for a critical review of the selection, measurement, and the use of social indicators. While we are obviously concerned about this problem as it relates to the social impact of the space program, in practice the starting point for consideration of the status of social indicators can as well be dictated by the general nature of the problem. This can readily be established by a casual thumbing of the Brookings Institution report. The range of possible effects of the space program which have been suggested is so great that it is impossible to think of any social phenomenon which has in the past been used as a basis for evaluation which can be presumed to be unrelated to the impact of space exploration.

As of now, there is neither possibility of, nor merit in, considering the space program as relevant to anything other than the full range of possible social indicators. We shall commission a critical review, looking at what in fact are cited as signs of trends in society, how they are measured, selective tendencies to use certain types of phenomena (say those which appear in numerical form), or selective tendencies in interpretation of ambiguous data, etc.

The problem of choosing and observing social indicators will not of course be limited to this one phase of our activities. On the general level, it relates to the question of national goals, particularly with respect to conversion of abstract goals into actual observations. It also relates to the question of feedback into organizations in that observations must be made in such a form that they can be transmitted, and so that the transmitted information in turn can be used relevantly to actions which might be taken. The issue of social indicators will also be a red thread running through the more specific studies. Community studies and studies of special groups, for example, would certainly suggest new social phenomena which should be considered in assessing the impact of the space program. So would the diffusion of space technology. It is expected that by the end of our grant our knowledge of *what* should be measured will have been helped equally by the frontal approach to the problem and by the more specific studies.

B. *The Space Program and National Goals*

The exploration of space has been defined as one of our national goals. The object of the NASA grant under which this project is operating might be stated as that of relating the activities associated with space exploration to other of our national goals; i.e., to what extent does space exploration help or hinder other national values. Existing studies of national goals, or of multiple objectives for organizations in general have generally suffered from two defects: (1) They have resulted in an enumeration of unquestionably desirable objectives without much guidance for trading off values when confronted with the inevitable conflicts among them. Even a priority listing is inadequate guidance, since the choice with which one is confronted is seldom that of one value *or* the other, but rather of some amount of one vs. some amount of the other. (2) National goals are usually stated at a sufficient level of abstraction so that there is a latitude of definitions. As a result, instead of becoming operational objectives they become battlegrounds (or weapons, depending on one's choice of imagery).

An essay is proposed which at a minimum will relate the problem of national goals to that of social indicators. That is to say, taking a usual list of national goals, this essay would pursue the question of what observations would have to be made in order to know whether we were approaching or retreating from the goals. Hopefully in this, or some further essay, present thinking on trading off among goals would be explored. Modern utility theory offers an approach *in principle* to this problem, and mathematical models such as linear programming are designed to handle multiple objectives. There remains the question as to whether such approaches are practical for issues as complex as those with which we are concerned. Attempts to employ highly complex, sophisticated models for decision making meet with peculiar resistance. Conversations with informed persons suggest that such computerized techniques make explicit the problem of trading off among values. It is our suspicion that many persons when forced to make such trade-offs explicitly are not only in conflict, but regard the activity as inherently immoral.

C. Feedback into Organizations

"Common sense" organizational theory assumes that "information is a good thing." A more sophisticated view is that this platitude must be modified with respect to the amount, kind, form, and source of information, the channels whereby it is communicated, and the place in the organization into which it is fed. In human organizations, knowledge of the consequences of one's actions can produce maladaptive responses ranging from rigid denial of the evidence to equally dangerous over-action. Most people and most human organizations are more complex than most machines, but even in machine systems, feedback can under certain circumstances be destructive.

The use of information in organizations is a widespread and long-standing activity of sufficient complexity and of sufficiently controversial a nature that we would not hope that a single essay or study would suddenly bring complete order to the problem area. However, there are two features of the feeding into NASA information of the social impact of the space program that may warrant explicit consideration. (1) The information about social consequences has to do with second-order consequences, concern for which will interfere with pursuit of the primary goal. In fact, even if there is no conflict between primary and secondary goals, the mere expenditure of time to think about the secondary goals constitutes a diversion of effort from the primary goal. Second-order consequences, to *some* groups, however, may prejudice pursuit of the primary goal. How can the system be organized so that it does not give a preferential advantage to those publics who can affect the survival of the space program? (2) The gathering and reporting of such information would put an "alien" group (presumably social scientists of a sort) in a position of power vis-à-vis the technical, scientific, and administrative personnel concerned with space exploration. This is true, even if no formal means of feedback into NASA is established. The doing and reporting of empirical research in public sources will

constitute a form of feedback. What will be the evolving relationship between those personnel whose responsibility is preferentially
to the primary goal, and those personnel more concerned with
second-order consequences?

Minimally a scholar sensitive to such organizational problems
should be asked to comment briefly on the distinctive sorts of problems and responses which might be anticipated. While these problems are accentuated in the space program, they are not in any
sense without precedent in other organized activities.

II. Devices of Anticipation

Here we are concerned with the question of understanding what
the social consequences of the space program *might* be. Second-
order consequences of actions may be desirable or undesirable. But,
if they can be anticipated, one can decide in advance whether or
not to try to avoid or to facilitate them. Attempts to anticipate consequences are in a sense "theoretical" exercises since they are dependent
upon relatively generalized notions about the *type* of consequences that flow from the *type* of action in which one is engaged.
Our intention is to explore some sources of such general ideas about
what type of consequences may result from an activity like the
space program, and how they may or may not be related to the
space program itself.

It is not our intention to add to the *volume* of the already extensive, informed speculation on the possible outcome of exploration of
space, but rather to look at some bodies of data and sources of ideas
that might serve as checks on the existing body of speculation.

A. *Historical Studies*

Historical situations analogous to space exploration will be exploited for possible generalizations which may aid in anticipation
of the impact of the space program. Examples of such studies are:

(1) *The social impact of technological change.* This calls for a
general essay by a scholar acquainted with a range of historical and
anthropological examples from which he can draw some applicable
generalizations. The focus of the study would be on whether or not
there is anything to learn from past experience as to whether or not
the social effects of technological innovation can be anticipated
and/or guided in a benign way. The very raising of this question
may pose something of a dilemma for the citizens of a liberal de-

mocracy. We know enough from anthropological literature to be aware of the possibility of a single technical innovation changing the entire structure of a society. For example, among the Tanala, there was a change from dry to wet rice culture. This brought about in sequence: (1) a change in patterns of land tenure; (2) a restructuring of family and kinship structure; (3) a complete change in political organization. In the light of this, a liberal society may have to plan and control in order to maintain its freedom.

(2) *Events which changed national self-concepts or man's image of himself.* Voyages of exploration and events such as the Copernican revolution have been alleged to have produced changes in the self-conception of nations or of mankind as a whole in a way similar to what may result from the exploration of space.

(3) *Studies of culture contact as analogues of anticipating extraterrestrial life.* In our view the question of whether or not there is extraterrestrial life is not so pressing as the fact that as we proceed on into space people will be concerned with this question, and, as happens so often, the anticipation may be more intense than the reality. Anthropological accounts of contacts between cultures, one of which would by some criterion seem "superior," might give us clues as to how people will handle the anticipation of possible contact with superior beings.

(4) *The impact of the invention of the microscope as a source of revelation of another form of life.* One of the issues of space exploration is that of discovering *sub*human but dangerous forms of life. The reaction to the discovery of microorganisms is a potential source of insight into reactions to a comparable event in space exploration.

B. *Nonhistorical Analogues*

Admittedly this is a catchall category, and possibly only a way of keeping the door open for other ideas we may have. An example of such an analogue is the public reaction to computers. This reaction becomes pertinent when one listens to how the general public responds to the notion that computers can "think." Careful attention to these reactions suggests that they may be based upon anxiety that man will be displaced in the phylogenetic scale from his top place as the only being that can "think." It is probable that for many people the ability to "think" is man's supreme capacity, and the basis on which their own self-concept is founded. As we proceed into the Space Age the importance of "thinking" and "intelligence" will probably increase as a measure of man's worthiness. In such a situation, the anticipation of encountering beings of superior intelligence (regardless of whether this anticipation is well founded or not) may become a source of increasing anxiety. Popular reactions to computers might be explored as a possible analogue to this anxiety.

It should be noted that the above paragraph is a flat contradiction to a generally held notion that exploration of space is or will be a source of anxiety to people of fundamentalist religious belief. The position implied in the above passage is that the threat is maximal for persons whose self-justification is man's ability to "think," and minimal for those persons whose self-justification is man's relationship to God.

It should also be noted for the record that the reactions of the general public to thinking machines discussed above bears no relation to the concerns of competent specialists actually working with computers. The "informed" public is worried by an entirely different set of problems.

C. *Studies of Fantasy*

Most "responsible" thinking is constrained by an adherence to the probable. Proper anticipation of consequences demands an adequate inventory of the possible.

Typical resources of fantasy pertinent to space exploration are science fiction and humor. Each reflects both an anticipation of the future, and some concerns of the present expressed in symbolic form. Any exploitation of these sources, within the limits of this grant, would not result in a systematic inventory of possible "consequences," but rather in a methodological review of how such sources might be used for anticipation. It would be injudicious at this point to rule out categorically any source of fantasy. Thus, while we are not prepared at this point to propose an exploitation of pathological fantasy, we would in principle be open to proposals to explore the fantasies of the mentally ill as sources of "anticipation." For example, it seems possible to obtain a historical series of responses to such fantasy provoking stimuli as the Rhorschach test and thereby learn: who was first "concerned" with space exploration, and which elements of space exploration first attracted attention. Obviously, this would fall into the category of basic research.

In using fantasy sources as devices of anticipation, both the impact of space exploration and of advanced technology may be studied.

D. *Substantive Studies*

(1) Anticipation of manpower demands. The exploration of devices of anticipation discussed to this point involved the consideration of analogues, both historical and nonhistorical, and the consideration of the use of "nonresponsible" imagination, i.e., fantasy. To round out the picture, there should also be some consideration of the use of responsible intelligence for anticipating consequences of space exploration.

The single space related problem which may presently be getting the most systematic attention from various quarters is that of anticipating manpower demands of the space program, both in the absolute sense, and in competition with other potential uses of manpower. There would be little point in the Academy Committee involving itself directly in the substantive problem of predicting manpower demands. However, the existence of the other efforts directed at this problem offers an opportunity for a methodological study of existing models and modes of predicting. The thrust of the Committee's efforts in this direction would be toward a critical evaluation of existing activities as a minimal goal, and the possible proposal of a substitute methodology as a maximal goal.

An interesting issue to which to attend would be the distinction between "anticipation" and "prediction" made at an earlier point in this document. Recent exposure to at least one attempt at "prediction" of future developments, commissioned by a government agency, suggests that more strict attention be drawn to the distinction between describing the most probable outcome and the most important foreseeable outcome. As a matter of fact, planning efforts may get into very serious difficulties unless specific probabilities and specific values are assigned to the full range of anticipatable outcomes.

(2) Management of expectations. It was indicated at the beginning of this work statement that the organization of it is convergent. Each successive proposal has generally been more concrete and more directly related to the social impact of the space program than the previous one. The present topic is at the point of transition between "devices of anticipation" and "detection of effects." It has to do with people's expectations of the space program; the yard-

sticks whereby they will judge specific events to be successes or failures; and the ways in which these expectations develop.

Terms such as "success" or "failure" of enterprises as complex as space exploration may be applied with abandon in the abstract. But the occurrence of specific events demonstrates that various publics have each their own set of expectations by which they judge the event to be a success or a failure. The history of the space program already offers adequate illustration of this point. Quite a few aborted missile flights were considered to be successes by the technical public because of the information which was gained from them. It is highly probable that a majority of the general public regards any flight which does not achieve full intended performance to be a complete failure.

Responsible management of so large a national enterprise poses an important problem for NASA officials as to how they should relate to the expectations of the various publics. The cumbersome term "relate to" is used, because there is in fact a two-way relationship. Public expectations are part of the world to which the NASA administration must respond, but these expectations are themselves a result of previous events in the space program.

The boundaries of this problem are hard to define, and a successful attack on it would be very much dependent on the imagination and the intellectual discipline of the man who undertook the task. What is proposed is an essay that draws on the existing history of the space program, relevant experiences in other areas, and foreseeable potential critical events in the space program. The essay would pose such questions as these: (1) What do we know in general about how relevant publics form their expectations of large enterprises such as the space program? How do they react to a violation of these expectations? (2) How would these questions be answered specifically for the American civilian space program? What range of actions is available to American officials to create realistic expectations, and to respond to violation of these expectations whether realistic or not?

An exhaustive treatment of the issue is beyond our scope. However, it is entirely within reason to expect some substantial progress on it. The work of Bauer and Furash, appended to the Brookings report,[1] already gives information on some of the expectations of

[1] See also Raymond A. Bauer: "Executives Probe Space," *Harvard Business Review,* 38 (September–October 1960), 6; "Keynes via the Backdoor," *Journal of Social Issues, 17,* No. 2 (1961), 50.

the business commmunity as of almost three years ago. The continuing work of Furash, described below, will give additional information on this topic, as well as offering evidence of trends over time.

III. Detection of Effects

Ultimately, the program of detection of effects would consist mainly, in a developed feedback system, of a continuing monitoring of important effects of space efforts with a reporting of these effects in appropriate terms to the appropriate agency. This activity would be accompanied by devices for early sensing of *unsuspected* effects. At this present early stage the problem requires a somewhat different emphasis. The activities proposed here are addressed toward three questions: (1) *Can* effects which have been guessed at be measured with adequacy? (2) *Can* procedures be devised for locating effects which have not been thought of? (3) Is it possible to segregate the effects of the space program from study of the effects of other factors in our society?

Studies proposed in this section will be different in form from those proposed above. In a few instances specific hypotheses will be tested, and it will be clear that the issue is explicitly one of whether or not a postulated effect can be measured. More often, however, we will propose exploration of some *area* of impact which seems either to be of practical importance, or which seems to offer some tactical advantage at this stage of investigation. The exploration of such areas will involve both the testing of specific hypotheses, and attempts at detecting unsuspected effects. While such studies will unquestionably produce findings of substantive interest, they are primarily *exercises* in the empirical study of the impact of the space program. They are to be evaluated in terms of whether or not they make possible and probable further and better studies along the same line, or demonstrate the undesirability and/or impracticality of such studies, or may also demonstrate the value of further study within a broader context than space.

Though these studies are intended to be primarily methodological, their orientation toward problems or problem areas prompts us to discuss them under topical headings rather than to attempt to organize them under general headings. The topics selected have been chosen to afford a variety of experiences, each of which should have general meaning in future efforts.

A. *Social Change in Space-Impacted Communities*[2]

Comparative studies are proposed in communities which have felt directly the impact of the space program and in similar communities which have not been so affected. Such findings would offer an opportunity for testing some hypotheses, and formulating additional ones, about "effects" of the space program that have not yet been anticipated or detected. Furthermore, while we do not for a moment harbor the belief that a present-day "local community" such as Huntsville, Alabama, or Route 128, Massachusetts, is a microcosm of tomorrow's Spage Age nation, the study of such a community may be a useful exercise prefatory to assessing the over-all societal impact of the space program.

An example of a broad hypothesis that might be tested is that the space program, by virtue of its emphasis upon technical training and knowledge, and the value it places on intelligence and intellectual skills, has created a new stratum in space-impacted communities. This stratum in turn, will develop, or will have developed its own distinctive way of life, and "ideology." Our hypothesis is that this new group will form an elite. But, equally important, it will have an effect upon the old high-prestige groups in the community as well as upon other groups of lesser prestige. Assuming that the general hypothesis seems plausible on preliminary examination, an attempt would be made to trace the effects of this development throughout the community.

While community studies offer an opportunity to explore such broad hypotheses as that stated above, they also offer, by virtue of the concentration of effort they permit, a special opportunity for locating totally unanticipated consequences. The systematic search for unanticipated consequences poses special problems. Naiveté, while it often intrudes itself where it is not wanted, is hard to come by deliberately. It takes effort to be deliberately naive, which is another way of stating the methodological truism that observations are always guided by some assumption whether explicit or implicit. One way to offset this fact is deliberately to incorporate random samplings of behavior and other cultural products for purposes of locating phenomena that our prior "wisdom" might bias us against looking for. Whether or not this latter course is pursued is subject to practical considerations of time and money.

---

[2] The proposal for studies of space-impacted communities is phrased as an outline model within which more limited studies will be initiated. This will become clear below.

The above scheme for comparative community studies could be carried out at varying levels of thoroughness, with varying expenditures of resources, and other varying time periods. What has been set forth is a set of objectives and procedures which might be approached in these varying fashions. *The extent to which we will be engaged in the pursuit of these objectives is presently moot.*

What is proposed specifically is a rapid overview of the existing knowledge of space-impacted communities, and of presently ongoing studies of such communities. At that point, a decision will be made on the best strategy for approaching this problem. Our present estimate is that our first serious approach to community studies would be from the point of view of special groups in the community. In effect, the community studies could at this stage be as well regarded as one of the studies of special groups outlined immediately below.

An example of a strategic group to study within space-impacted communities would be the high school students and teachers. It is expected that the major value and status changes hypothesized above would be reflected acutely in the high school system. This system might be approached via students, teachers, guidance specialists, school board members, etc.

B. *Impact of the Space Program Upon Special Groups*

The community studies mentioned above are aimed at tracing the impact of the civilian space program on spatially contiguous groups. It is equally desirable to trace the relationship of the space program to functionally important groups which may be physically dispersed, i.e., educational institutions, scientists, businessmen, etc. In an effort to capitalize on prior work, the Committee made an early decision to study the impact of the civilian space program on the business community.

Goals of the study of the relationship of the business community to the NASA program are: (1) to explore the range of impacts and reactions; (2) to attempt to measure changes in effects over time, capitalizing on base-line data gathered two years ago; (3) to develop techniques for anticipating future reactions under specifiable contingent conditions.

Even what seems to be the simplest of these objectives — to measure change over time — poses some real difficulties. It would appear at first glance that the task would be merely one of repeating the observations made at the earlier period and comparing the results for the two time periods. However, many of the issues have changed over the two-year period, and therefore the same questions cannot always be asked. Furthermore, our own understanding of the issues and the methods of studying them have improved. This faces us with the perennial question of how to improve one's methods and yet gather comparable data over time.

Once more attention is called to the overlap between the studies of special groups, and the approach to community studies indicated above.

C. *The Diffusion of Space Technology*

One of the most widely discussed second-order consequences of the space program is the diffusion of space generated technology into the civilian economy. Of all second-order consequences, this one is perhaps also regarded with most optimism by supporters of the space program. Part of the problem is the over-all economic contribution of such technology; another problem is the process of transfer itself. Macroeconomic studies of the impact of military and space technology are under way in several quarters. Therefore, even if they were within the Committee's capacities, we would be disposed to do other studies which would make a more distinctive contribution.

In concentrating on the process of transfer, we are making the explicit assumption that the key problem in the contribution of space technology to economic growth will be that of the translation and adoption of available technology.[3] The supply of new knowledge and technology will be very large and generally available. The economy which grows fastest will be that which can *use* generally available knowledge. Here there is already evidence that the Japanese experience with "imitation" may give them a lead in what will be a crucial skill in the future.[4]

Study of the process of transfer has an additional advantage. Usually, in speaking of "consequences" and "impact" we tacitly treat the entity being affected as though it were passive in the process, a victim or beneficiary that has little to do with its own fate. This tacit assumption is especially prevalent in the area of technological innovation where there is a tendency to assume that a new technical development of potential value will almost automatically be adopted. Failure of adoption is generally treated with

[3] Cf. Robert A. Solo, "Gearing Military R & D to Economic Growth," *Harvard Business Review, 40,* No. 6 (November–December 1962), 49.

[4] For our immediate purposes, it is necessary to assume that economic growth is *per se* a desirable goal, although at other times we may question this.

more impatience than intent to understand. In earlier passages of this document, we raised the question of whether or not the consequences of technological innovation can in fact be anticipated and/or controlled. Study of the process of diffusion of space technology gives us an opportunity to explore a situation in which the opposite circumstance seems more likely; i.e., being "influenced" may be dependent on a good deal of activity and cooperation on the part of the person or organization being affected. That is to say, "influence" is in fact often a *reciprocal* process rather than a unilateral one. The incorporation of any new element into an organization is a complex process systematically related to structure and functioning of the organization.

(1) *A case study of a single restricted innovation, tracing as far as possible all presently identifiable developments which seem to flow from this innovation.* The purpose of the case study would be to improve our understanding of the range of effects we may expect to find as a guide to what should be studied in the future. Attention would be directed to the twin questions of *what* was adopted, under what circumstances? Was it the actual product or process, or was it generalized knowledge, or did the product or process suggest an idea that generated an entirely new innovation? How did it happen?

(2) *Development of a general approach to technological transfer, using space technology as an example wherever possible.* This study would concentrate on characteristics of the adopting organization that effect the probability of adoption of innovation, especially innovation that originates outside the organization. Our decision to focus on the organization *per se* is a reaction to the fact that approaches to adoption of innovation have usually been concerned with general economic variables or such broad organizational characteristics as size of firm, industry patterns, etc., on the one hand, or with quite strictly technical considerations, on the other hand. In the usual discussion of technical aspects of adoption of innovation, "organizational" considerations seldom extend beyond the organization of the R & D unit itself. An "advanced" treatment considers the role of R & D in the organization. As one moves toward more general treatment of characteristics of the organization, the amount of concern drops off rapidly. Hence, our effort would be to develop an understanding of the area that seems most neglected in the over-all problem of technological diffusion.

From a point of view of our general mission, we would be asking with respect to a given type of "consequence," the circumstances under which it occurs, and what can be made to help it occur where and when it seems desirable.

D. *The Level of Public Reaction to Space Exploration*

Public reaction to many important states of affairs has posed a difficult problem, both of interpretation and policy. For example, in the period after World War II, people showed a surprising apparent lack of concern over the dangers of the atomic bomb. This was interpreted variously as "genuine apathy" or as "repressed deep anxiety." Over certain ranges of events the difference in interpretation makes no practical difference. In a crisis the difference could be crucial. If apparent apathy were in fact a systematic psychological defense against deep anxiety, there could indeed be circumstances in which behavior would be sharply discontinuous with what had gone before.

There is reason to think seriously of the level of public reaction to space exploration. A decade ago stories about "flying saucers" caused great excitement. Two decades ago a radio broadcast simulating a Martian invasion caused panic. Yet today, with space exploration an imminent reality, the public's manifest level of anxiety over the implications of space exploration seem if anything to have decreased. Considering the fact that progress in space exploration is tightly linked to the threat of missile attack, there is a double reason for questioning this apparent lack of increased anxiety.[5]

The task of discriminating between apathy and repressed anxiety is difficult but not impossible. Accordingly, we would regard with favor a study which would relate public opinion poll data with other psychological data for the purpose of determining whether or not these two sources of information produce a discrepancy that would imply repressed anxiety.

Apart from the interesting methodological aspects of this problem, it also raises the possibility of bias in the sources which generate hypotheses about the impact of the space program. Intellectuals, particularly social scientists, have speculated widely on the potentially disturbing implications of space exploration for fundamentalist religious groups. On the other hand, we have suggested in a preceding portion of this report that exploration of

---

[5] This passage is not written in ignorance of the fact that some groups in society are indeed *very* anxious and active over the threat of thermonuclear war. Our concern here, however, is with the apparent lack of worry about "flying saucers," "men from Mars," etc.

space might produce its greatest anxiety among those groups who put an especially high value on intelligence as man's most distinctive attribute. The key assumption in this argument is that anticipation of discovering more intelligent beings in space would constitute a threat to this latter group, while there is little that would of necessity threaten a religious person's relationship to the Deity. If this counterhypothesis should be supported, we would have a valuable illustration of how anticipation of impact of the space program may be biased by the point of view of the analyst. Demonstration of such a bias may serve as a prophylaxis against bland acceptance of what appears to be "self-evident." It should also underscore the value of empirical checking of such propositions if they are of crucial importance.

The issue of anxiety, for example, could in fact prove to be of extreme importance in the Space Age. Presumably, the Space Age will bring greater prestige to men of intellect, and elevate intellectuals to a new elite status. In recent times, intellectuals have been among the least religious groups in society and — almost by definition — the group which most finds its justification in its intelligence. This would mean that the crucial group in the carrying out of space exploration would be the group most threatened by such activities. Among possible outcomes might be either a certain degree of paralysis, or the adoption of a consoling religious belief.

Index